Country Crafts

Apple
Chutney
October 1990

Mincemeat
November 1990

Emmerdale

Country Crafts

Joy Gammon

YORKSHIRE
TELEVISION

B⊞XTREE

SPECIAL NOTE

The craft items in this book are graded
according to difficulty:

*   easy, suitable for beginners;
* *   average, for people with some
     experience or who would like to try
     something more difficult;
* * *   difficult, for the experienced or those
     who enjoy a challenge.

Measurements and quantities are given in
metric and imperial. Follow one or the other;
do not mix them.

For Maggie Heath

First published in the UK in 1991
by BOXTREE LIMITED, 36 Tavistock Street,
London WC2E 7PB

1 3 5 7 9 10 8 6 4 2

© (text) Joy Gammon 1991
© (photographs and illustrations) Boxtree Limited 1991

Emmerdale is a registered Trademark of
Yorkshire Television Limited.

Design by Maggie Aldred
Project editor Helen Douglas-Cooper
Photography by Stephen Morris
Illustrations by Diana Groves

Set in Goudy
Typeset by Tradespools, Frome
Colour origination by Fotographics, Hong Kong
Printed and bound in Hong Kong by Wing King Tong

A catalogue record for this book is available from the
British Library.

ISBN 1 85283 108 1

CONTENTS

Introduction

The Yorkshire Dales, where *Emmerdale* is filmed, is a very beautiful area of England where traditional values are strong. Over the centuries the people have chosen, or have had chosen for them, hard ways of life, hill farming, shepherding, working in mills or even down mines. Because of this reliance upon the old livings, the old ways live on, and among them a love of making do for oneself. This includes a kind of scorn for bought-in food and consequently a strong tradition of home cooking, especially baking. There is a strong sense of thrift, and so any item of clothing that can be made rather than bought is valued the higher.

Because of this preference for relying upon one's own hand to survive, to live, to enjoy and even to embellish everything around, there is a natural survival of the traditional crafts and skills in the Dales. People who do not live there, look at the crafts and see in them a symbol of these values, and want to try the skills for themselves.

There is a tremendous satisfaction in looking at something and saying 'I made that'. This book is designed to help anyone who would like to achieve that kind of satisfaction, whether in cooking, needlecrafts or home decoration. The things in the book are also chosen to give great pleasure in themselves, whether they are made as gifts or to keep.

These recipes and crafts often use natural materials, things which are sometimes neglected, often cheap and occasionally free. If you are a newcomer to any kind of craft or to country cooking, I hope that you will enjoy this introduction to the ways of Emmerdale people. If you already make all kinds of things of your own, there is such a variety here that I am sure you will find something new to try, or an old friend to return to. I would like to give my grateful thanks to all the Dales-women who helped me with anecdotes, advice and ideas, and I hope that this book will give everyone a taste of Emmerdale.

JOY GAMMON

COUNTRY SKILLS

The garden provides an abundance of riches to be used in traditional ways around the home. Add seasonal fragrance to your rooms with pot-pourri. Use the flavour and scent of herbs to make refreshing teas and relaxing sleep pillows. Blend the delicate hues of dried flowers in posies and decorative wreaths.

Pot-pourri

Add a splash of colour and delicate fragrance to a room with a bowl of pot-pourri made from a selection of spring, summer, autumn or Christmas mixtures.

In the past, scented petals and herbs were used in great profusion in wealthy homes to cover other, much less pleasant, smells. Today, these mixtures are used because they are attractive: pot-pourri is both very pretty and has a delicious smell. It is usually placed in china bowls or special pot-pourri containers which add to its charm. A room which contains such a bowl is especially welcoming, and evokes a bygone and more gracious age. The choice of plants for pot-pourri is a personal one, and part of the fun is in picking and mixing all sorts of variations, because pot-pourri means just that – a mixture. The method described here is for dry pot-pourri, and consists of mixing different combinations of dry ingredients which smell and look good with essential oils.

DRYING YOUR OWN MATERIAL

If you have a garden full of roses or herbs, you may like to dry your own pot-pourri ingredients. This is very satisfying and gives a comfortable feeling of being able to save those summer days to enjoy right through the winter.

Always pick material when it is already dry, for example in the middle of a sunny day when there is no chance of dew or rain. Collect flowers and petals just on the point of opening and do not include stems, as it is the petals which have the perfume. Lavender flowers should be picked just after their flowering peak, and herb leaves should be picked with care to protect the rest of the plant. Use secateurs or strong sharp scissors to cut stems, and make sure that cuts are clean and do not damage the remaining plant. Apart from the fact that this kind of damage leaves the garden looking devastated, it is bad for the plants and encourages disease.

A delicate spring pot-pourri (top) in soft greens and pinks, made from rose-buds, marigold, senna leaves and dried daisy-like heads; a rich summer mixture (bottom) of lavender and blue mallowflowers, decorated with pink rose-buds.

Spread the plant material to dry in an airy, but not too hot, place. Direct sunshine will cause the petals to discolour and will dissipate their scent. Ideally air should be allowed to circulate around the material, so if you can, place them on thin fabric (such as an old net curtain) stretched over a frame. There is no need to make a frame; old picture frames could be used, or the net fabric could be temporarily stretched over a clothes airer and either tied on or attached with drawing pins if the airer is not particularly special.

Suitable flowers and plants to grow and pick for pot-pourri include:

anemones	hibiscus	roses
box	iris	scented geraniums
carnations	lavender	stocks
cornflowers	marigolds	sunflowers
eucalyptus	pinks	violets

A wide range of pot-pourri materials are available commercially, and the ingredients in the following recipes were bought ready dried. Several items have to be bought anyway, unless you have an exceedingly exotic garden; for example, few people grow cinnamon or ginger. Scented oils and essences make delicious-smelling additions which enrich the natural fragrances of the material, and these can also be bought. Both individual perfumes, such as orange, and special pot-pourri mixtures are available. They are usually bought in a small bottle, rather like the vanilla and lemon essences used for cooking, and sometimes the bottle has a glass dropper fixed into the lid. If not, it may be useful to buy a dropper of this kind from a chemist, as it is easier to control the application of the essence, which is very strong, in this way.

It is a good idea to add a fixative to the mixture as it helps the perfume to last and sometimes has a scent of its own. Powdered orris-root is the easiest to obtain, and should be available from chemists and herbalists.

Finally, because pot-pourri is also decorative, the recipes have been given a seasonal theme and one or two ingredients which have no scent have been added to each to make the final mixture look

even more attractive. Any type of decorative material can be added at this stage, including artificial items. For example, very small silk flowers could be added to floral pot-pourri, or wax red berries and tiny glass baubles added to the Christmas mixture.

General method

Quantities are a matter of taste and preference. Pot-pourri usually contains a bulky main ingredient, such as rose petals, to which complementary scents are added. The mixing process is very simple. All ingredients must, of course, be dried, as any fresh material would go off and cause all the remaining material to spoil.

Mix together all the larger plant ingredients in a large bowl or on a tray. Do not use a wooden bowl or tray as it will retain the perfumes for a very long time and affect everything else that it is used for.

Sprinkle on the spices and oils, using the glass dropper if preferred, and mix well.

Sprinkle on the fixative at the rate of about $2^{1}/_{2}$ g per 500 g ($^{1}/_{8}$ oz per lb) of dry material.

Mix well daily for several days to blend the scents, then divide into bowls and add the decorative ingredients, if any, to the top.

Note: Pot-pourri is often placed in polythene bags until used, for example when you wish to give it as a present. This does retain the scents, but polythene should not be used for long-term storage as condensation can gather, causing the material to deteriorate.

SEASONAL POT-POURRI *

Ingredients for these pot-pourri were chosen for their seasonal colours, and many are dried and kept from other seasons. The pleasure is in experimenting with what you have available and the perfumes that you like.

SPRING
Bulk ingredients: rose buds, clover flowers, senna leaves, marigold flowers, lavender (a fairly small quantity as lavender is strongly scented)

Additions: optional scented pot-pourri oil, powdered orris-root
Decoration: any dried daisy-like flower

SUMMER
Bulk ingredients: rosebuds, blue mallow flowers, lavender
Additions: powdered orris-root
Decoration: whole lavender flowerheads

AUTUMN
Bulk ingredients: marigold flowers, tarragon, bear berry leaves
Additions: optional scented pot-pourri oil, small quantity ground ginger, powdered orris-root
Decoration: red hibiscus flowers

WINTER
Bulk ingredients: bear berry leaves, red hibiscus, thyme
Additions: small quantity ground cinnamon, small quantity ground nutmeg, cloves, orange oil, cinnamon oil, powdered orris-root
Decoration: pine cones, pieces of cinnamon stick

DISPLAYING POT-POURRI

Special jars or containers can be bought which have pierced lids or sides, allowing the scent to escape. These can be very beautiful and many people collect them. They are also very practical, as the covered material retains its character and is effective for longer. However, the pot-pourri cannot be seen, and as it is attractive in its own right, it is equally successful displayed in simple bowls or pots around the room. Again, do not use wooden containers as they will become impregnated with the scent, and unlined baskets are not successful because the material falls through the spaces in the wicker and, because the basket is so open to the air, the pot-pourri will give off a very strong perfume for only a short time. It is better not to put bowls of pot-pourri in direct sunshine as, again, they will dispel their perfume too fast, and the material will fade and discolour.

Pot-pourri makes a delightful gift, with or without a container, and can be made to suit the recipient. It can be packed temporarily in a polythene bag or in a fine fabric bag tied with ribbon.

Subtle golds and browns are used in the autumn pot-pourri (top), marigold flowers, tarragon and bear berry leaves; the spicy winter mixture (bottom) is made from bear berry leaves, hibiscus and thyme, decorated with pine cones and sticks of cinnamon.

Herbs

Herbs are easy to grow, take up little space, and can be used in many different ways that make the most of their wonderful flavours and soothing properties.

Herbs are surrounded by mystery and legend. Man has grown them ever since he began to cultivate the soil, and has used them for magic, medicine and cookery. They are still used to add interesting flavours to food, or to perfume linen, or to make a soothing or medicinal drink.

Herb-growing is especially satisfying because it is easy, and provides an ideal way for beginners and children to learn the mysteries of garden cultivation. The plants are varied and interesting, very attractive when growing, and can be used in all sorts of culinary and other household ways. An added bonus is the wealth of folklore which surrounds the growing and use of herbs, and which can become a study in itself. Indeed, many books have been written about this aspect of herbs, and the collection of this kind of fascinating information helps to make the growing of herbs a most absorbing pastime.

THE HERB GARDEN

Herb gardens can be quite small, as the average cook does not use vast quantities of any herb. Most herbs are easy to grow, although some, like parsley, can take a very long time to germinate. Choose a well-drained, sunny corner as many herbs are Mediterranean in origin and appreciate such conditions. Choose the herbs that you prefer, and either grow them yourself from seed or buy them ready to plant. Such container-grown herbs thrive from the start as their roots have not been disturbed, and the plant can be cut for cooking or eating as soon as it is in the ground without the risk of setting it back or using it before it is ready.

If you use larger quantities of any herb, such as chives or parsley, it can be grown in rows in the vegetable garden. Some herbs are very decorative in their own right, and plants such as sage can be reared in the herbacious border. Sage has delicate, pretty flowers which also enhance a flower garden, and many other edible herbs such as rosemary and mint will flower if allowed to. Take care with mint. It is attractive and comes in many forms, such as peppermint, as well as the more familiar kind. However, it is also self-propagating, and once it has settled in it pops up all over the place to the detriment of your more sensitive plants.

Some herbs, such as rosemary and bay, are really shrubs and need to be treated as such. They are happy in large tubs or pots, with plenty of watering, and can look beautiful in a small patio garden where space is at a premium. Because these shrub-herbs can grow to a considerable size, it is worth planning carefully where to put them, as they could go in either the herbaceous border or the herb garden. Do not let them shade other herbs, or make the rest of the garden inaccessible. Large plants towards the front of a garden also spoil the look of it and render smaller and more retiring herbs invisible.

Herbs were often planted in patterns. They have such varied foliage, and only small numbers of plants are needed, so they can be arranged in a circle, a grid, or any other pattern. One very pretty idea is to plant the herbs within an old wagon-wheel, each section filled with a different variety of plant.

If you have no garden, then herbs love well-tended window-boxes, and are even happy in decorative pots on a sunny window-sill. They are ideal for the inexperienced gardener, being easy to grow, and are, of course, indispensable for the dedicated cook. The pleasure of having your favourite varieties available at all times makes herb-growing very satisfying.

HARVESTING AND DRYING HERBS

Herbs taste and smell the way they do because they contain their own special oils. These are volatile and the aim when drying herbs is to remove the water content without allowing

Fresh herbs can be grown in a traditional herb garden or on the window-sill, and are easy to look after. They can be used to add interesting flavours to cooking, and to make herb vinegars and soothing herb teas.

much, if any, of the oil to escape.

Herb flavours are at their strongest just before flowering, so this is the time to pick them. Cut with stems as long as possible early on a warm day so that the foliage is dry but the perfume has not been destroyed by the sun (see page 10).

Ideally the herbs should be dried quickly and in the dark with as much air around them as possible. Use a frame covered with netting if you can, as for pot-pourri (see page 10), or place them on the racks of the airing-cupboard. Traditionally the bunches were hung from the kitchen ceiling, cut stems upwards, and in a Yorkshire farmhouse kitchen there were always hooks in the beams for hams, puddings and herbs. Another old-fashioned gadget was the overhead drying-rack (a wooden airer suspended over the cooking range which could be lowered for the clothes to be hung on it on washday). If you have one of these still, but don't use it for your washing, then herbs can be hung from it to dry. In any kitchen, modern or old, bunches of herbs hung up to dry look good, smell good, and are just where you need them when you want to cut a sprig or two for cooking.

Be careful when handling the drying herbs or the leaves will break off. When they are dry, though, you can rub the leaves gently off the stems and store them in labelled airtight jars. These should either be pottery jars or, if glass, kept in a closed cupboard, as light discolours and denatures the herb material. Do not store herbs for any length of time in bags. Paper bags allow the flavour to escape, and any moisture which becomes trapped in a polythene bag will cause the herbs to go mouldy and deteriorate.

Herbs are best laid out on a rack to dry. Once dry, they can be used to make wonderfully scented sleep pillows and lavender bags.

Template herb vinegar label

HERB VINEGARS

These are extra tasty vinegars made by steeping herbs in vinegar and used to give additional subtle flavours to any dish which would normally be eaten with a vinegar dressing. Herbs such as mint, dill, basil and tarragon can be used.

Although raspberries are not strictly a herb, a recipe for raspberry vinegar is included because it is so delicious and unusual. Raspberry vinegar of the kind described here appears on Yorkshire tables to be eaten as a sweet sauce with the ubiquitous Yorkshire pudding, and is not to be confused with the kind of raspberry vinegar used in modern cookery for salads and sauces.

TARRAGON VINEGAR *

PREPARATION TIME: AT LEAST 4 DAYS

▲

sprigs of fresh tarragon
white wine vinegar

▼

Put the sprigs of tarragon into glass bottles or jars and cover completely with white wine vinegar. Tightly close the lids on the jars. Leave for about four days, then strain into glass bottles, cork firmly and store.

If a stronger flavour is required, repeat the process with further fresh tarragon.

They say in the Yorkshire Dales that a family that can grow good parsley is a family in which the wife wears the trousers.

Vinegars infused with herbs such as tarragon and basil add a special flavour to cooking, and also make excellent gifts.

RASPBERRY VINEGAR *

PREPARATION TIME: A TOTAL OF ABOUT 3 DAYS

▲

450 g (1 lb) raspberries
malt vinegar
sugar

▼

Put the raspberries in a basin and cover with the vinegar. Leave for 2 days, stirring occasionally. Strain through a jelly bag (see below) or clean cloth. Measure into a pan and add 350 g (12 oz) of sugar for every 600 ml (1 pt) of juice. Bring to the boil, stirring until the sugar is dissolved. Boil for 20 minutes. Allow to cool, then transfer into glass bottles and cap firmly.

This vinegar has a delicious sweet and sour flavour and is eaten in Yorkshire with the Yorkshire Pudding (see page 72) which makes up the first course of a North Country dinner. Some people even use it as a cough mixture.

Cooking time approximately 30 minutes.

Jelly bags are so called because their primary use is to strain the juice from fruit which has been cooked to a pulp and which will be used to make jelly jams. Jelly bags can be bought in various sizes, or a fine clean cloth, preferably white, can be used. The cloth will stain and so should be kept for this purpose. A jelly bag usually has tapes to hang it with, but a cloth can be tied with string which is then used to hang it up, or opposite corners can be tied in pairs to create a bag effect and these knots used to suspend the bag. If no suitable hook or hanger is available, the bag or cloth can be hung from the legs of an upturned stool. When pouring the pulp into the bag or cloth for straining, first place the bag in the bowl which is to receive the juice and pour the pulp into the bag and bowl together. Then carefully lift the bag from the bowl, leaving some of the juice behind and allowing the remainder to begin to flow into the bowl. Do not squeeze the juice through the bag or cloth unless a recipe specifically states that this should be done, as it will make the juice cloudy and less attractive in appearance.

HERB TEAS*

Herbal teas are made in just the same way as Indian or China tea and are very popular in Europe, and especially in France where they are called tisanes. They are regarded as medicinal and soothing, and many people prefer them to the stimulating effect of caffeine in ordinary tea.

Warm an ordinary pottery teapot, and spoon into it about three teaspoonfuls of the chosen herb for every $^1/_2$ litre (1 pt) of tea to be made. Use a little less of dried herbs as they are stronger. Pour on the boiling water and leave for a few minutes. Pour and strain, and add sugar to taste, but not milk.

Try making herb tea with any of the following:
camomile – for soothing tiredness; also good for colds.
lime blossom – for soothing tiredness.
marjoram – for curing headaches and for good health.
mint – to aid digestion.
sage – to sooth stomach ache.
nettle tops – to sooth rheumatism.

Herb teas can now be bought in many shops so that it is possible to experiment with all the various flavours. They can even be purchased in separate teabags so that all the different kinds can be tried, both for flavour and effect.
Note: Be very careful that any plant material which you use, especially if you have picked it in the wild, is safe and clean, and that you have identified it correctly. Also, although these herb teas are fun, and may give some genuine relief, they are no substitute for a doctor's opinion if you are really ill.

SLEEP PILLOW*

A pretty, lacy sleep pillow filled with a relaxing herb mixture; and lavender bags, which are ideal for scenting drawers of linen.

This is a small pillow or cushion filled with a mixture of dried herbs, which has proved to be an effective cure for sleeplessness. It consists of a double pillow: an inner large gauze sachet filled with the herb mixture, and an outer decorative cover. The herb sachet goes inside the outer pil-

20

*An infusion of Pimpernel
and Rosemary, used as a
scalp rub, will help to keep
a fine head of hair, and
sweet sage tea aids
a bad memory.*

low and is placed alongside the sleeper's head. It is not used as an actual pillow, as it would not be very comfortable!

The herb mixture can be bought ready made, or you can make your own by mixing together dried rose petals (see the instructions for pot-pourri on pages 10–12) with small quantities of cloves and dried mint. Or make a mixture of rosemary, woodruff and lavender. Another lovely sleep pillow mixture is made from peppermint, lemon verbena, lemon thyme, and camomile. The traditional filling is perhaps the easiest: just fill the pillow with hops for a blissful night's sleep.

The sleep pillow can be any size, depending upon your preference and possibly upon the amount of sleep herb mixture available to fill it. The most attractive and easiest shape to make is probably the traditional pillow shape, a rectangle whose shorter side is about two-thirds of the length of its longer side. You will need sufficient gauze and decorative fabric for two rectangles of each in the required size, together with ribbon and lace for trimming. You will also need a quantity of sleep herb as described above for the filling.

Cut the gauze into two matching pieces of the preferred shape and size, allowing an extra 1 cm (¹/₂ in) seam allowance all round. Also cut the decorative fabric into two matching pieces in the same shape as the gauze but allowing an extra 2 cm (1 in) all round: 1 cm (¹/₂ in) all round to give a larger cover than the gauze, and a 1 cm (¹/₂ in) seam allowance.

Using a sewing-machine, or backstitching by hand, with right sides facing, stitch together the two matching pieces of gauze around three sides, about 1 cm (¹/₂ in) in from the edges. Turn the bag right side out, fill with the herb mixture and then, folding in the raw edges neatly, oversew the fourth side to close the pillow.

Using a sewing-machine, or backstitching by hand, with right sides facing, stitch together the two matching pieces of decorative fabric around three sides about 1 cm (¹/₂ in) in from the edges to form the outer cover. Turn the cover right side out.

The decorative cover can be trimmed in a variety of ways with ribbon and laces. Lace can be

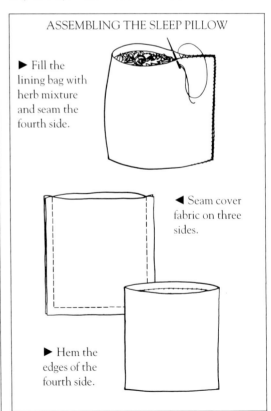

ASSEMBLING THE SLEEP PILLOW

▶ Fill the lining bag with herb mixture and seam the fourth side.

◀ Seam cover fabric on three sides.

▶ Hem the edges of the fourth side.

stitched around the edges of the pillow, and gathered around the corners, or it could be layered across the pillow to create an all-over frilled effect. Bows of ribbon could be made or bought and stitched on at the corners, or as preferred. Other decorative ribbons, ribbon rosebuds or appliqué motifs could be stitched on, but do avoid beads, buttons or any hard decoration which could be uncomfortable.

Place the herb bag inside the decorative cover. Fold over raw edges of fourth side, and either sew together or insert a zip, to close.

These bags can be made with any herbal or pot-pourri mixture, and can be decorated in any way you wish. They make delightful presents, but are not suitable for very small children, who might eat the trims or even the contents.

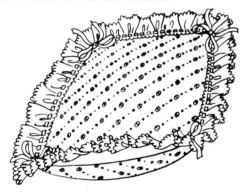

▲ Place herb pad inside cover and decorate with frills, ribbons, etc.

FRAGRANT SACHETS AND LAVENDER BAGS*

Lavender is one of the most attractive herbs, not just because it smells so good, but because it looks beautiful growing in the garden, attracts the bees, which love it, and is very easy to dry and use. These sachets could be made with any herb, but are usually made with lavender, which keeps its scent well. They are ideal for placing amongst linen and clothes. They can be made from any fine fabric that will allow the scent to escape, and can be decorated with anything that you have in the way of ribbons and trims.

▲

Dried lavender
Small quantity of fine fabric
Ribbons and trims

▼

Take two matching small squares or rectangles of the fabric and, with right sides facing, seam them together around three sides. Hem the top raw edge of the bag either by machine or hand, then turn the bag right side out.

Part fill with lavender, then tie firmly shut with a piece of ribbon. For extra safety attach the ribbon with a few firm oversewn stitches on the seam. Stitch on ribbon bows, roses or pieces of lace.

On Yorkshire farms there is always considerable demand for the beastings, or colostrum, the first milk which a cow gives after calving. It is far too rich to be sold for bottling, and is often a strong creamy colour, even yellow, and very thick. It makes the most wonderful custards and cheesecakes, setting on its own when heated, and is delicious simply heated with sugar and sultanas.

Baskets

By decorating baskets with fabric and trimmings, or fresh or dried flowers, and filling them with a selection of useful little items, you can create pretty and unusual gifts for special occasions.

Baskets are one of the oldest kinds of receptacle and many ancient cultures used them for carrying everything from cheese to bricks, and made them out of everything from feathers to copra. They are popular today for their beauty and nostalgic charm and also for their great practicality. They are tough, light and hard-wearing and come in all sizes from egg cups to chairs. You may be lucky enough to have a basket shop in your area. If not, try gift shops or the kind of shop which sells everything for the flower-arranger. Either way, you will find a treasure trove of baskets in an almost endless range of sizes, shapes, colours and materials.

They make such an attractive gift, even empty, but lined and decorated with fabric trims, bows and posies, and filled with presents for special occasions, they make unique gifts. There are many other ideas which can be added to the ones given here to create the perfect present, which, even when the contents are gone, provides a lasting souvenir.

A wonderful present for a youngster, or for someone retiring, would be a fishing basket filled with suitable tackle items, and any hamper could be filled with hobby equipment for any of a score of activities. If someone is retiring or leaving work, why not get everyone to buy something from a list and assemble the gifts, all separately wrapped, in a suitable basket. Think about a painting basket, with brushes, palette and tubes of colour, or a games basket with chess, Scrabble and jigsaws. A toy basket full of separately parcelled toys would be a delight for a child at Christmas, and a log basket full of all sorts of essentials for the household would be wonderful for a young couple setting up home for the first time.

A basket makes an ideal receptacle for a fresh flower arrangement, as the flowers can be displayed to good effect to show off their different colours and textures.

FRESH FLOWER BASKET * *

If you would like to have a special table centre of cut flowers, or to give someone flowers presented in an unusual way, you can create a basket of flowers which, because they are in water, will last as long as they would if in a vase. Baskets and flowers always go together well, but do be careful to choose appropriately. Fresh flowers can be quite heavy and baskets are light, so avoid irises and delphiniums unless your basket is enormous. Go for lighter, shorter blooms, and remember to top up the water often.

Seasonal flowers look especially good in baskets, with their suggestion of being picked fresh into the baskets in the meadow, and spring prim-roses or cowslips in a small, pale basket are charming. Choose a larger, lavish basket for summer short-stemmed roses, and a tall basket for the longer-stemmed autumn flowers such as chrysan-themums. Try to echo the flower shape in the shape of your basket, putting, for example, spikes and grasses in tall vertical containers, and round flowers in lower, round baskets.

▲

A suitable basket; here, a pale round basket approximately 18 cm (7 in) across (see the note about handles below).
A water container to fit out of sight within the basket. A piece of flower-arranging block suitable for fresh flowers to fit the water container.
Assorted fresh flowers and foliage as chosen. Florists' wire.

▼

Place the block into the container and carefully and gently wire the block into the container, and both into the basket, as invisibly as possible, using the technique described on page 26.

Add water as required until the block is completely soaked and there is excess water in the container. Trim flowers and arrange into the block, noting the following special points:

Try to decide clearly where flowers should go and then leave them there, moving them from

point to point as little as possible. This is because holes made in the block remain, and a great deal of indecision will cause it to disintegrate.

Keep stems fairly short to avoid wilting and also to keep the whole arrangement more compact and stable.

If the basket has a handle which you will wish to use when the arrangement is complete, plan the height of the flowers before you arrange them, making sure they will clear the handle sufficiently. This is why many fresh flower baskets have a proportionately high handle, and you may wish to choose one of this type.

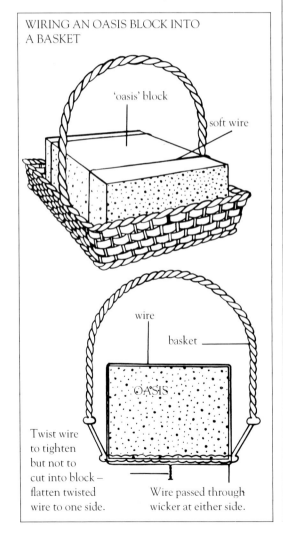

WIRING AN OASIS BLOCK INTO A BASKET

'oasis' block

soft wire

wire

basket

OASIS

Twist wire to tighten but not to cut into block – flatten twisted wire to one side.

Wire passed through wicker at either side.

A lasting display of dried flowers. An arrangement in a basket is an ideal way to give flowers.

DRIED FLOWER BASKET**

For general notes on growing and preparing dried flowers, see page 10.

A basket of dried flowers is a lasting gift, and can provide decoration all the year round. Use seasonal colours or even silver and gold, and choose a basket which is simple enough to complement the flowers without being too decorative in itself. All flowers are beautiful, but they look especially good in colour groups, so try baskets all of red or yellow flowers or an assortment of pastels. Dried flowers are an excellent gift.

▲

A suitable basket; here, a dark oval basket approximately 18 cm (7 in) across
A piece of flower-arranging block suitable for dried flowers (oasis) smaller than the base of the basket
Assorted dried flowers as chosen
Florists wire

▼

Carefully and gently wire the block as invisibly as possible into the bottom of the basket. Cut two lengths of wire, either with wire clippers or with old scissors, each of which is long enough to pass around the whole block and through the basket with about 5 cm (2 in) to spare at each end. Pass each wire over the block and thread the ends through the bottom of the basket, twisting them together under the basket to hold the block in place. Take care not to pull the wire too tight, as it will begin to cut into the block, which may then start to break up.

Arrange the flowers as desired and stick the stems into the block. General notes on arranging dried flowers are given on page 36–9, which provide a useful guide to the best method for this.

Choose flowers that are not 'top heavy', as a basket is itself not very heavy, and any weighty seed-heads or tall flower spikes could cause it to fall over. Try to use fairly dense material which will hide the block and, because a basket is a light and delicate container, use airy and delicate plant material to achieve a balanced effect.

BABY BASKET*

One of the most startling discoveries following the arrival of a new baby is the large number of things that you need to carry around with you, both from room to room at home, and when out and about. A basket is ideal for this as it stands on its own and holds everything upright. It will even last long enough for the small recipient to use again one day for the next generation. So make a useful and pretty present by filling a simple basket with some of the most useful baby things.

▲

Deep, flat-bottomed basket.
Ribbons and silk flowers for decoration. (To make this basket suitable for either sex, or even twins, aquamarine and lilac were used rather than the more obvious pink and blue).
A range of baby needs, such as: baby towel, baby oil, face cloth, baby talc, baby shampoo, cotton-wool balls, petroleum jelly, nappy pins, baby brush and comb, and (essential) a teddy bear suitable for a new baby.

▼

Note: Make sure that everything you give to a new baby is especially made for the purpose and very safe in every way.

Decorate the basket with flowers and ribbons and arrange the contents attractively in it. Small bunches of flowers can be made and attached as described for the Special Occasion Basket on page 30, and ribbons can be threaded through the wicker of the basket or around the handle, and tied in bows. Such decoration should be removed long before the baby is old enough to remove and eat any of it.

WORK BASKET**

Baskets are ideal as containers for all your needlecraft items, but they do need to be lined to prevent all the smaller items falling through the holes. A padded lid is a sensible convention too, as a place to put pins and needles while you are working. Such a basket makes a lovely gift, and looks so much better in use than the alternative of

a collection of polythene carrier bags. It is also an excellent gift for a child.

▲

1 lidded basket; the round basket used for this project is approximately 20 cm (8 in) in diameter and 14 cm (5¹/₂ in) deep.
Sufficient wadding to make a cushion for the lid. Here, 20 cm/8 in diameter circle of wadding material.
Sufficient bright cotton fabric to line the basket and make both sides of the cushion padding the lid. Here, a piece 85 × 41 cm (31 × 16¹/₂ in).
4 self-covering buttons.
For the contents: a range of needlecraft materials.
Especially useful would be embroidery scissors, measuring-tape, needles, thimble, stitch-holder, pins, knitting needles, silks, ribbons, fabric and sewing cottons.

▼

Cut a 20 cm (8 in) diameter circle of wadding to line the lid. If you are using a different size or shape of basket, cut the wadding to fit inside the lid (see diagram, page 30).

To line the sides of the basket, cut a rectangle of fabric with its longer edge equal in length to the internal circumference of the basket, and the shorter edge equal to the depth, adding a 1 cm (¹/₂ in) seam allowance all round. In this case the circumference is 79 cm (30 in) and the depth 14 cm (5¹/₂ in), so a piece of fabric 81 × 16 cm (31 × 6¹/₂ in) is needed (see diagram, p. 30).

Cut out a circle of fabric to line the base of the basket, allowing a 1 cm (¹/₂ in) seam allowance all round. In this case, the base of the basket has an internal diameter of 20 cm (8 in), plus 1 cm (¹/₂ in) all round, giving a diameter of 22 cm (9 in) (see p. 30).

Cut two further circles which will form the cover for the padded cushion in the lid. To allow for the thickness of the wadding, these need to be about 10 per cent larger than the piece of wadding. To cover a 20 cm (8 in) diameter piece of wadding, and allowing for 1 cm (¹/₂ in) seam allowance all round, two 25 cm (10 in) diameter circles of fabric were cut.

Baskets are very versatile and can be adapted to all sorts of uses, such as a work basket for holding sewing things, or a baby basket filled with gifts.

FABRIC PIECES FOR A CYLINDRICAL BASKET WITH LID

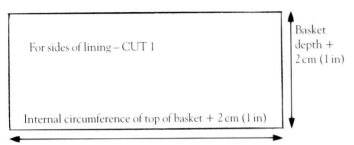

For sides of lining – CUT 1

Basket depth + 2 cm (1 in)

Internal circumference of top of basket + 2 cm (1 in)

For cushion pad in lid

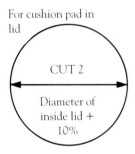

CUT 2

Diameter of inside lid + 10%

For lining base

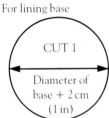

CUT 1

Diameter of base + 2 cm (1 in)

Baskets are perfect for picnics (top), or can be used to create an unusual gift to mark a special occasion or anniversary.

To make up the lining, first fold the rectangle of fabric in half with right sides facing, and join the short sides together either by machine or backstitching by hand. Make sure that the circumference of the cylinder of fabric matches the circumference of the inside of the top of the basket.

With right sides together, seam in the circular base fabric to the bottom edge of the lining, by hand or machine as before. If the basket tapers inwards towards the bottom, you may need to gather the bottom edge of the side lining slightly to fit the base before seaming evenly, using the 1 cm (¹/₂ in) seam allowance.

Turn the 1 cm (¹/₂ in) seam allowance to the wrong side all around the top edge of the lining and tack down.

Attach the top edge of the lining to the inside top edge of the basket. Depending on the type of basket, the best way will probably be to work small groups of stitches at close intervals to catch the lining to the wicker. Remove the tacking stitches.

With right sides facing, and using the 1 cm (¹/₂ in) seam allowance, join together the two circles of fabric for the lid by machine or backstitching by hand, leaving a section of the seam open. Turn right sides out and insert the wadding, ensuring that it is flat and even. Close the remainder of the seam by oversewing.

Cover the buttons with fabric following the instructions given with them.

Position the buttons in a regular pattern on the lid cushion, and stitch them firmly through the wadding and through the wicker lid, so quilting the cushion and fixing it firmly to the lid.

Add the contents.

SPECIAL OCCASION BASKET*

For an anniversary, or any very special celebration, what better present than a bottle of champagne and two beautiful flutes from which to drink it? Better still, if they are packed in a decorated basket, it too can be kept and used for years as a reminder of a happy day.

This gift was made for a Silver Anniversary, so glittering paper and silvery decoration were used, but you can use whatever is appropriate to the occasion. Anniversaries are not only celebrated by people, of course, and special baskets could be made for churches and societies, buildings and schools. A basket of cups and saucers for the church hall, a basket of camping accessories for the cubs, a basket of books for a school, are other possibilities.

▲

*Flat basket large enough to take a bottle.
Decorative paper or packing; such as white tissue under irridescent transparent cellophane.
Dried flowers and artificial flower material, some sprayed silver.
Decorative ribbon.
Florists wire.
1 bottle champagne.
2 champagne flutes.*

▼

Make decorative bunches of the floral material and attach to the basket. Such flower sprays can be bought if preferred, or simply made by bunching a small and attractive selection of dried and/or silk flowers and binding them together with soft florists wire. The same type of wire can then be used to attach the spray or sprays to the handle or to the basket itself, taking care to leave no sharp ends that might hurt someone.

Line the basket with decorative paper or packing, and put in the champagne and glasses. The base of a bottle of champagne is much heavier than the neck, so take care when lifting the basket that it is balanced or the bottle could roll out.

FOOD BASKET*

Baskets are the traditional way of carrying a picnic; where would Ratty, Toad and Moley have been without their basket of goodies? You could take a hamper with all the plates and glasses strapped into the lid, but this is a much simpler version, perhaps just for two to take up the Dales on a lovely summer day? The basket here would make a lovely present especially as a simple cloth and napkins have been made to go in it. If the basket is to be a gift, make the food very special. The basket itself, the cloth and napkins and the pretty picnicware will serve as lasting reminders of a lovely day and a successful present.

▲

*Fairly large, strong basket.
Square of gingham for cloth, approx 113 ×
113 cm (45 × 45 in)
Two squares of contrast cotton fabric for
napkins, each about 40 × 40 cm (16 × 16 in)
Two pieces of ribbon to tie the napkins.
Picnicware for two. It will be more practical if
these are unbreakable.
A selection of special food such as bread sticks,
cheeses, home-made biscuits, sweets, caviar,
bottled meats and savouries. Wine, juice or
bottled water.*

▼

First, make the cloth and napkins. Turn a double hem to the wrong side of the gingham cloth and stitch down either by machine or by hand-hemming. Hem the edges of the cotton squares in the same way. Double each serviette, roll into a cylinder, and tie with a ribbon. Line the basket with the gingham cloth and arrange the food appetizingly in it.

MOSES BASKET*

This is a very spectacular present for an expectant Mum or for a very new baby. It can only be used when a baby is very small, because as soon as the baby can pull himself up using the basket sides it may tip over. However, the basket will always be useful. This idea could also be used to dress the

A Moses basket, dressed with a broderie anglais trim, is ideal for a new baby.

CUTTING OUT WADDING
AND FABRIC

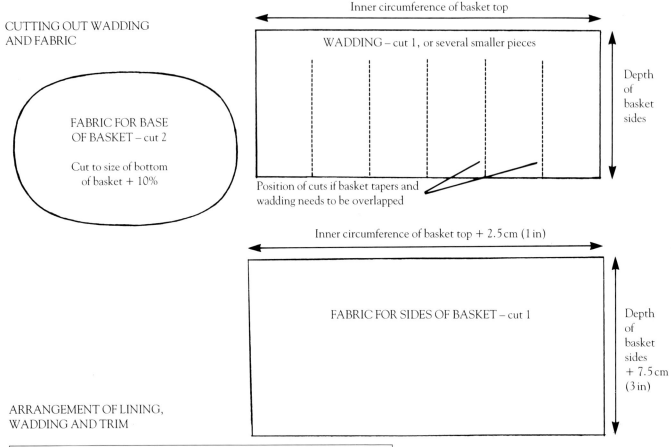

Inner circumference of basket top

WADDING – cut 1, or several smaller pieces

Depth
of
basket
sides

Position of cuts if basket tapers and
wadding needs to be overlapped

FABRIC FOR BASE
OF BASKET – cut 2

Cut to size of bottom
of basket + 10%

Inner circumference of basket top + 2.5 cm (1 in)

FABRIC FOR SIDES OF BASKET – cut 1

Depth
of
basket
sides
+ 7.5 cm
(3 in)

ARRANGEMENT OF LINING,
WADDING AND TRIM

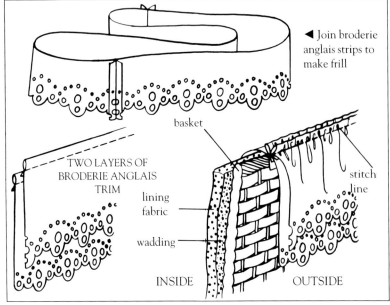

◄ Join broderie
anglais strips to
make frill

TWO LAYERS OF
BRODERIE ANGLAIS
TRIM

basket

lining
fabric

wadding

stitch
line

INSIDE

OUTSIDE

other classic improvized cot – a deep drawer from
an old-fashioned chest-of-drawers. Do take care,
when making anything for a baby, that all the materials are completely safe and non-toxic.

▲

A very large laundry or similar basket.
Sufficient plain white cotton fabric to line the
sides of the basket, and to cover the bottom of
the basket twice over.
Sufficient nylon wadding to line the sides of
the basket.
A clean old blanket or blanket material to line
the bottom of the basket.
Sufficient broderie anglais edging to go around
the top edge of the basket twice.

▼

Cut the wadding into suitable shapes to line the
sides only of the basket. If the basket tapers

towards the bottom make vertical cuts in the wadding and overlap the edges of these at the bottom so that the wadding fits snugly round the sides. It is not necessary to attach the wadding as the fabric which is going to cover it holds it in place, but it could be stitched by hand to the wicker of the basket at intervals.

Measure the circumference of the basket around the top edge, and add on a 2.5 cm (1 in) hem allowance; measure the depth of the basket and add on a 7.5 cm (3 in) seam allowance. Cut a piece of the white fabric to fit the measurements. With the right sides facing, join the two shorter ends of this piece of material along the seam allowance either by machine or backstitching by hand.

Cut two strips of broderie anglais edging to the same length as the top edge of the piece of fabric for the side lining of the basket. For each strip, place right sides together and join their ends, by machine or backstitching by hand, to form two circular frills. These are going to be sewn to the top edge of the lining, so take care to make sure that they match exactly in length. Turn a small hem to the wrong side along the top edge of the lining and catch it down by stitching on, by hand or machine, the two frills of broderie anglais edging so that they fall outside the top edge of the basket when the lining is in place. Put the lining in place with this edging hanging outside the basket and the lining covering the wadding on the inner sides.

From the blanket or blanket fabric cut several layers the size of the bottom of the basket, or fold the blanket to fit.

From the remaining white cotton, cut two pieces to fit the shape of the bottom of the basket, plus a generous hem and filling allowance of about 10 per cent. With right sides together, seam these pieces by hand or machine around three sides to form a cover for the blanket pieces. Turn right side out and stitch a small hem by hand or machine along the raw edges of the open ends so that the cover can easily be removed for washing. Arrange the blanket pieces or folded blanket pad inside the cover, and place in the bottom of the

basket where it will hold the bottom of the lining in place.

Safety note: NEVER LINE THE BOTTOM OF A BABY'S COT WITH SOFT WADDING, OR ANY KIND OF SPRINGY OR FLUFFY MATERIAL. Always make babies' mattresses firm and do not give them a pillow, as they can suffocate in soft materials. If in doubt, choose a basket which is large enough to take a carry-cot mattress. You could also ask your local health visitor or midwife for advice about all the materials before you begin.

The following counting
rhyme, used by shepherds,
occurs in the Celtic
languages, and there are
Cumberland versions. But it
has also been used for
generations in the Yorkshire Dales.

●

Yan, tyan, tethera, methera, pimp;
Sethera, lethera, hovera,
dovera, dick;
Yan-a-dick, tyan-a-dick, tethera-a-
dick, methera-a-dick, bumfit;
Yan-a-bumfit, tyan-a-bumfit,
tethera-a-bumfit, methera-a-
bumfit, giggot.

Dried and pressed flowers

Dried or pressed flowers can be used in many different ways from delicate and simple decorative objects such as greetings cards and straw hats to large, elaborate arrangements.

Dried flowers are becoming more and more fashionable as decoration for the house, partly because they are very beautiful, and partly because they are permanent without being artificial. They can be arranged in spheres, rings or bunches, and can have corn, seed heads and other materials mixed in with them. Many people have seen the beautiful traditional continental decorations made in European countries, and have been inspired to have a go for themselves so making the decorative use of dried flower material more and more popular.

These arrangements can have all kinds of moods. They can evoke the farmhouse through generous bunches placed in jugs or hanging from a beam and corns and grasses in profusion, or they can be more formal, with large arrangements of branches and leaves of the kind one might expect on a great carved sideboard. More usually, however, because of the amount of space available, they are small and simple, such as posies, small baskets and hanging spheres; or, for the more sophisticated, there are bay tree and wreath arrangements.

There are three main sources of dried flowers: those which you pick in the wild and dry yourself, dried garden flowers, and bought material much of which comes from more exotic plants that many people are not able to grow. With greater experience it is possible to experiment with all kinds of attractive material to find out what works and fits in best with your chosen style.

Wild plants suitable for drying for decorative use include grasses and reeds, docks, hemlock, mullein, wild clematis (old man's beard), thistles, bracken and yarrow. Tree material such as small branches, cones and seed pods are useful too.

Many garden flowers can be satisfactorily dried. Especially suitable are lavender, nicotiana (tobacco), buddlieia (butterfly bush), various ferns, chinese lanterns, hebe, golden rod, poppies, hydrangeas, honesty and teazles. There are also the so-called everlasting flowers, including varieties such as helichrysum, which are almost dry on the plant.

It is possible to buy all of these varieties in specialist shops or at garden centres. You will also find exotic twigs and barks, unusual grasses and reeds, and a very wide variety of flowers. Many bought dried flowers are dyed, which can add wonderful and subtle colours to an arrangement, although very bright unnatural colours should be avoided unless you want an artificial effect.

Material which you pick and dry yourself should be harvested and dried in the same way as pot-pourri (see page 10), so that the plant material is dried without destroying its character. In addition, however, because the flowers here are to be used intact, great care must be taken not to crush them. Gather the material in small bunches, and do not tie them too tightly. Take care not to lay too many bunches on top of one another, and not to cover them with any heavy material. Always handle with care to keep any fragile petals as perfect as possible.

GENERAL INSTRUCTIONS FOR MAKING A DRIED FLOWER ARRANGEMENT

BASED ON AN OASIS BLOCK

▲

A container, if appropriate
A block of flower-arranging material, such as oasis. Make sure that it is the correct kind for dried flowers
Fine florists' wire
Plant material
Secateurs and either wire-cutters or an old pair of scissors
Special materials (see project instructions).

▼

Prepare the container by placing the block of oasis inside, wiring it in firmly and as invisibly as possible. Because water is not necessary for dried

Spheres of dried flowers can be adapted to hang on a ribbon. Alternatively, they can imitate the classic bay-tree shape.

flowers, containers such as baskets can be used. Or small holes can be pierced with an awl in the bottom of a plastic container so that lengths of wire can be threaded through these, over the block, and then twisted together under the container to hold the block in place. Take care not to pull the wire too tight or it will cut into the block, and ensure that any twisted ends of wire are safely out of the way to avoid damage to fingers.

Cover the block with a layer of one kind of plant material, cutting pieces of the length you require, and choosing in particular any multi-branched plant to give a background layer of colour against which you can make your arrangement. Choose pieces with a strong stem and push the stem well into the block to hold it firm. Try to be decisive about the positioning of material – if pieces are removed and replaced too often the block will break up.

Now arrange the rest of the material to create the effect you require. It is preferable to work with one kind of flower or material at a time, distributing pieces evenly about the arrangement and then adding a similar distribution of another kind of flower until the required effect is achieved. Some materials can simply be cut to length and pushed into the block as they are, but you will find that there are some kinds of flower which have too

WIRING A FLOWER STEM

Insert one end of soft wire into the back of the flower head. Wind firmly around available stem.

weak a stem. To arrange these, cut the stem short and firmly bind the end with wire, leaving a long piece of wire free to replace the cut stem. This wire can then be pushed into the block. This technique also enables flowers to be bent into the position which you require, using the flexibility of the wire.

Continue in this way until the oasis block is covered and invisible, and you have the effect you wish.

Material such as poppy heads and cones can also be sprayed, for example with gold or silver for a festive effect, and any arrangement looks much

ASSEMBLING A DRIED FLOWER ARRANGEMENT

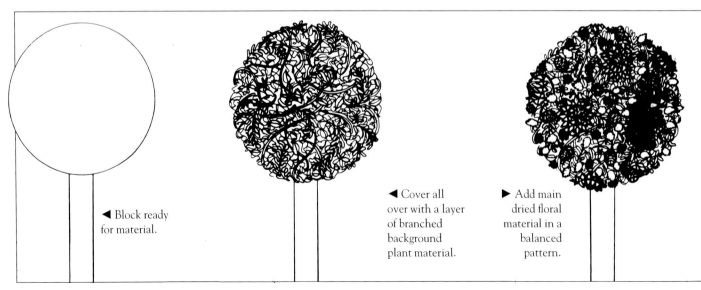

◀ Block ready for material.

◀ Cover all over with a layer of branched background plant material.

▶ Add main dried floral material in a balanced pattern.

more successful if a colour family is chosen and the same kinds of colours used throughout. For example, a pastel and pinks arrangement looks rather odd if bright red is introduced, although it always pays to experiment, and some of the best effects can be achieved by breaking the rules.

HANGING BALL OF DRIED FLOWERS**

These spheres are a traditional European decoration, especially in Germany, but they are becoming increasingly popular everywhere. They make a beautiful gift, and are ideal for the small house or flat as they do not take up much room. Those made here are in bright yellows and oranges.

▲

Basic materials (see page 36).
A ball of oasis.
Ribbon for hanging.
One piece of strong wire, about 1¹/₂ times the
diameter of the ball in length.

▼

Make a small flat loop at one end of the strong wire and then pass the wire straight through the ball of oasis so that it rests on this flat loop, which will take its weight. Make a hanging loop in the other end of the wire.

◄ Complete
with non-plant
material.

Cover the ball with the dried flowers as described in the instructions above. Because this ball is covered all over with flowers, it becomes impossible to hold. You can begin with the ball placed in the top of a jug or similar container until most of it is covered. To cover the remainder, spear the ball on a strong knitting needle, and place the needle in a bottle to support it. You can then get at the remaining surface. When the ball is complete this needle can simply be removed, and the hole which it made is not visible through the flowers.

Make a ribbon loop through the hanging loop of the wire at the top of the ball from which to suspend it.

DRIED FLOWER TREE**

This classic bay tree shape is very decorative. It can be made to any size as long as you have enough dried plant material. It could be made in green, as an imitation tree, but subtle pinks and blues are preferred here for a more delicate effect.

▲

Basic materials (see page 36).
Sphere of oasis.
Extra oasis blocks or small spare pieces of oasis.
Flower-pot type of container.
Suitable thick stick for a trunk. A natural
straight branch was used here, but a length of
broomstick could also be used, a garden cane or
something similar, depending upon scale.
Clean gravel, similar to the kind used in a fish
tank.

▼

Pack the flower-pot container about three-quarters full with the extra blocks of oasis, cutting them to fit if necessary. Then firmly push the stick into this oasis to form the trunk, ensuring that it is exactly vertical. Now push the top of the stick into the oasis sphere so that the oasis is centred on the stick.

Cover the ball with dried flowers as described above.

Top up the pot with the gravel to cover the oasis blocks supporting the trunk.

SUMMER DOOR WREATH*

This bought-cane wreath is so decorative that it was a pity to cover it up, so a spray of summery harvest plants and flowers with corn stems were added to enhance the warm colours of the cane.

▲

Basic materials (see page 36).
Cane wreath.

▼

Arrange the plant material into a decorative spray, lengthening any stems with wire if required, as described on page 38. Wire the spray together, and then wire it firmly but as invisibly as possible to the wreath. Flowers with wired stems can be positioned so that their heads cover any visible wire.

CHRISTMAS DOOR WREATH**

A Christmas door wreath can provide a lovely welcome to your house.

▲

Basic materials (see page 36).
A wreath-shaped container.
Blocks of oasis cut to fit into the container.
Very small red and silver baubles on wire stems.

▼

As invisibly as possible, wire the oasis into the container in the same way as described on pages 36–8, then cover with dried flowers as described in the main instructions.

Add the red and silver baubles to decorate.

Right: A cane wreath is attractive on its own, and can just be decorated with a spray of summer flowers and corn stems. Opposite: A wreath for Christmas in the seasonal colours of greens, reds and silver will provide a warm welcome on a chilly day.

DECORATED BOATER *

▲

*Miniature straw hat, available from
handicraft shops
90 cm (1 yd) each of very narrow satin ribbon in
contrasting colours
90 cm (1 yd) of 2 cm (1 in) wide satin ribbon
Assorted sprays of small dried flowers,
flower-heads
Needle and cotton
Saucer of transparent rubber-based glue
Sharp knife
Double-sided sellotape
Scissors*

▼

Place the small flower sprays at intervals around the hat brim, and sew the stalks firmly in place. Using a sharp knife, trim the bottom of the flower-heads so that they are flat. Cover them with a generous helping of glue and place them firmly in position around the brim, taking care to co-ordinate the colours as you go. Leave to dry overnight.

To make streamers, cut 25 cm (10 in) strips of each narrow ribbon colour, fold double and sew onto the back of the hat. Make a small bow from the same three colours and sew on, fanning out the ribbons.

To make the hanging ribbon, cut approximately 30 cm (12 in) of the wide ribbon and attach one end to the hat brim with double-sided sellotape, and fold the other end over to make a 10 cm (4 in) loop. Make two smaller loops from the remainder and sew on at right angles. Make two bows from the narrow ribbon and stick on with double-sided sellotape to hide the stitches.

POSIES **

The Victorians gave posies as keepsakes; some of them even contained hidden messages because the flowers had different meanings and lovers knew the code. Posies are becoming increasingly popular for weddings, too, often being made from silk flowers so that the bride and her attendants can keep them for ever. Traditional posies have a central flower surrounded by concentric rings of other flowers or decorative plant material, finished with a frill which was sometimes lace but more usually, then as now, pierced paper of the kind used for doilies, which are ideal for the purpose. The stem was bound with ribbon or tape.

▲

*Basic materials (see page 36).
Paper doily.
Ribbon or tape to bind the stems.*

▼

Take the central flower and arrange the other flowers or material around it in concentric rings. Unless the material has thin, flexible stems, you may find it easier to wire short stems as explained on page 38. As each ring of flowers is completed, bind the stems with fine wire to give a firm foundation for the next ring of flowers. When the posy is the size you want, make a hole in the centre of the paper doily and pass the complete bundle of stems through it. Fold and gather the doily and wire it to the top of the stems. Cover the wire and stem bundle by binding with tape or ribbon; secure free end by stitching or glueing.

PRESSED FLOWERS

Another way of preserving and drying flowers is to press them. Again, this was a sentimental Victorian habit, flowers which brought back memories being pressed in prayer books or favourite novels. However, this is not a good way to press flowers, and will also stain the book.

Choose delicate, fine flowers for pressing, not necessarily small, but not too thick, and not with fat juicy heads or stems. Place them on a sheet of flat absorbent white paper, such as blotting paper, ensuring that they are flat and smoothing out any creases or folds. The paper in its turn should be on a flat, smooth surface. Do not use paper with an embossed pattern as this will transfer to the petals, which are very delicate. Place another layer of the absorbent paper on top of the flowers and then cover with a large flat weight; books are ideal. The flowers will take at least a week, and should be left alone during this time as, until they are dry, they will tend to stick and split.

Dried flowers can be used to decorate a straw boater to stunning effect. They can also be made up into Victorian posies, either to be carried or used as table decoration.

After about a week, transfer the flowers to fresh paper and re-press in the same way for a further week. This helps to remove extra moisture while retaining the colour.

Finished pressed flowers are very frail and need to be stuck to a backing. They can be collected in a scrap book, or made into decorative items like the bookmark and card illustrated here. If you have many flowers you can make elaborate pictures.

PRESSED FLOWER CARD AND BOOKMARK**

Pressed flowers are meant for books, so here is a way of doing this without spoiling the pages. Both the card and bookmark are very easy to make and would be a thoughtful gift or greeting for someone special. You can use a blank greetings card bought from a craft shop, or you could make your own.

▲

Thin cardboard, or blank greetings cards.
Fabric glue.
Small sheet of glass or plastic.
Transparent adhesive plastic sheet (the kind sold for covering books).
Small bow of yarn.

▼

Cut the card to shape if required, either as a bookmark, or a greetings card if you are making your own card. The edges of the card need to be very straight or it will not close or stand satisfactorily. Remember, too, to decorate it so that it will open in the correct direction, that is, like a book, unless it is particularly required that it should open differently. If you are buying an envelope for the card, make the card to fit a standard envelope size, as it can be very hard to find an unusual size envelope at a later date.

Arrange the flowers on the card in a pleasing pattern, and stick them down. The easiest way to do this is to spread a thin layer of fabric glue onto a sheet of glass or plastic, carefully pick up the flower (tweezers can be used for this), and place it on the glue so that the underside is covered with a thin coat. Carefully lift the flower from the glue and place it in position on the card. Place the fin-

ished glued picture under a sheet of polythene and then lightly press until the glue is dry. Polythene peels away from glue much more easily than paper if any glue has escaped around the flowers.

Decorate the picture if you wish; for example, a bouquet arrangement can have a small yarn bow on the stems. Such decorations can also be stuck on with fabric glue.

Bookmarks, or any other items which are going to be handled regularly, should also be covered very carefully with self-adhesive plastic sheet. Cut a piece exactly the size of the card and remove the protective backing. Position it at the bottom of the card, and slowly and carefully press it down as you work up the card, squeezing out air bubbles as you go.

PRESSED FLOWER FRAME**

▲

Photograph frame with detachable back
Window mount (you can buy these separately)
Pressed flowers and leaves
Tweezers
Transparent rubber-based glue
Cocktail sticks
Double-sided sellotape

▼

Work out the position of your photograph within the mount, and carefully mark it.

Make sure that you have enough pressed flower material as specimens are easily damaged when handled. Position flowers and leaves around the mount using tweezers.

When you are happy with the design, fix in position by lifting up the first flower, smearing a tiny amount of glue onto the mount with a cocktail stick and placing the flower back exactly into position. Allow to dry overnight.

Remove any excess glue by gently rubbing it away, and carefully slide the mount into the frame, making sure that the edges do not catch.

Place strips of sellotape onto the back of the mount. Peel off backing paper and lower onto the photograph in position. Press firmly together, and replace the back of the frame.

Delicate pressed flowers are ideal for decorating objects such as bookmarks and picture frames; they also provide an easy way to make attractive greetings cards.

Stencilling

Stencilling is an easy skill to master, and provides you with an original way to transform a room, or to decorate a piece of furniture, fabric or even a biscuit tin, using a pretty or humorous pattern.

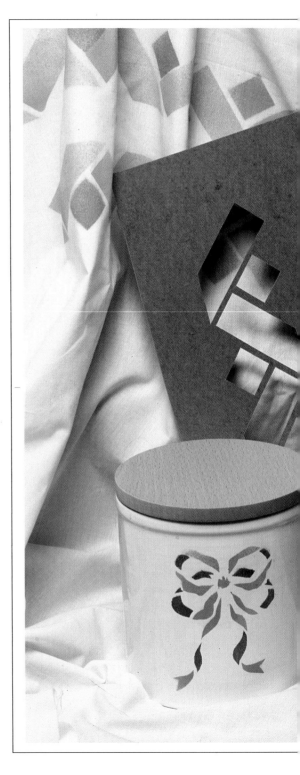

This old technique for producing unique decorations on walls, furniture and household items of all sorts has become very fashionable and popular. Stencilling is fun, and the results are very attractive. No special artistic talent is needed, just patience and care, although great creativity can be brought to bear with both spectacular and subtle results.

A simple room can be transformed into something very special and personal through the use of stencils. Depending upon the designs and the way in which they are used, stencils can add sophistication, detail or humour. An ideal way to experiment would be to stencil a frieze around a child's room. Stencilled decoration can also alter the apparent scale of a room: ceilings can be lowered, rooms widened and brightened, walls given more interest, and detail brought to eye level.

If the idea of stencilling on this scale is daunting, the technique can be used for all kinds of smaller things such as boxes and tins, greetings cards, parcels, and fabric items such as T-shirts and scarves.

CHOOSING A DESIGN

Typical designs suitable for interiors are given here, but designs can be taken from anywhere. Some people who have the talent design their own stencils from scratch, but many prefer to use or copy motifs and ideas from other sources. Whether using these sources for inspiration or to copy, there are endless possibilities. Think of natural forms like leaves and flowers, and, especially for children, animals. They will need to be stylized and simplified for stencils, so look at how this has been done on things like wallpaper

Stencils can be used to decorate a variety of household objects.

and tiles. Or you can choose a style which appeals to you, such as Victorian, Art Deco, Art Nouveau, or Egyptian, and look in books about the period, taking motifs from objects and illustrations. There are even books of sheets of designs from different periods which are specifically designed for use in this way. You can have enlarged photocopies made at specialist copy shops, and use them to trace off a stencil design, perhaps in a series of sizes.

The stencil itself is made up of a series of shaped holes through which the colour is applied. The pieces of stencil between the holes should not be too thin or they will pull and tear as you work, and you will need to bear this in mind when preparing your design. For example, the solid shape of a maple leaf is much more appropriate for a stencil than is a fern.

MATERIALS

Special oiled stencil board is cheap and durable. It is available in a variety of weights; try to get the thinner version, which is easier to cut. Stencils can also be cut from sheets of acetate. This has the advantage of being transparent so that when you are positioning the stencil on the wall, you can see

A YORKSHIREMAN'S ADVICE TO HIS SON
See all, 'ear all, say nowt,
Eat all, sup all, pay nowt,
An' if ivver tha dus owt fo' nowt, do
it fo' thissen.

what you have done already. Care is needed when cutting an intricate design in acetate, as it can tear easily. Board and acetate are available from craft shops and specialist stencilling suppliers.

Stencil brushes are cropped, round, solid brushes, and it is advisable to buy the best that you can afford. Choose a size appropriate to the size of the motif; have at least two and use them for different colours while you are working, to keep the colours clean and clear. Sponges can be used instead of a brush. It is easy to control the application of the paint using a sponge, in order to create a soft, subtle effect, or create shading. The colour can be built up in layers for a stronger effect.

Acrylic paints are the most suitable. Other paints can be used, but acrylics have been found to be the most suitable and long lasting.

Aerosol spray paints can be used, especially for larger areas. Take care to ask for the kind of paint appropriate to the surface on which the stencil is to be made.

You will also need paper for drawing the original design; tracing paper for transferring it to the board; a good-quality, strong craft knife; masking tape for holding tracing paper in place and also for correcting any mistakes; a flat, smooth work-top on which to cut the stencils (these can be bought from craft shops, or improvised using a wooden board or a sheet of glass with its edges bound with masking tape to prevent injury). If you are using aerosol paints it may be necessary to have to hand some cellulose thinners to remove any mistakes.

STENCILLING TECHNIQUE

Choose your design and check that it is the right size. You can alter it by reducing or enlarging it on a photocopier. If you are using stencil board, trace the design onto the stencil board, holding the tracing paper in position with small lengths of masking tape. If you are using acetate, place it over the design and fix it in position with masking tape. Trace off the design using a felt-tip pen or drawing pen designed for use on acetate. At this stage make sure that the shapes in the design are simple and clear, and that the connecting bars of the stencil make a cohesive pattern. Indicate clearly the pieces to be cut away by shading them in or marking them in some way.

Using the craft knife, cut out the holes in the stencil as neatly as possible, trying to cut each line with one stroke rather than 'chewing' at it and making a jagged edge. It is essential that the craft knife is sharp, so change the blade regularly. Be very careful while using it because craft knives are ideal for cutting people too.

If you make a mistake, stick pieces of masking tape over the mistake on both sides of the board or acetate and recut through the tape.

Decide on the exact position of the stencil on the wall or object to be stencilled. Use measuring-tapes, set squares and if necessary a plumb line on walls, and rulers and set squares on smaller objects and furniture, to do this as accurately as possible. It is surprising how far out of true walls can be, so it is worth taking care at this stage. The best way to find the middle of any rectangular area is to run two threads diagonally from corner to corner and mark the point at which they cross. Mark walls with a small pencil cross or chalk line. With an all-over design, it is probably best to begin in the centre and allow the design to run out at the edges.

Make sure that the surface to be stencilled is smooth, clean and free from dust or damp.

Attach the stencil in position. This can be done using lengths of masking tape at the corners; or even better, re-useable spray adhesive of the kind used by display artists (this can be bought from art shops), which holds the stencil in place all over.

If applying the paint with a brush or sponge, place the colours in the separate compartments of a palette, dip the brush/sponge in the paint and then, working with a generous rag in the other hand, almost dry the brush/sponge on the rag. Apply the paint by dabbing rather than stroking it onto the surface until the colour builds up to the required thickness. If using more than one colour, use a separate brush or sponge for each.

If using spray paint, mask the surrounding area

The beautiful Rowan Tree, or Mountain Ash, has several virtues. There are the umbels of creamy white flowers, and the bunches of red berries which make delicious jelly, or which, if you leave them on the delicate branches, will attract all kinds of birds. The Rowan Tree is also well known for keeping away witches.

with newspaper applied with masking tape and hold a sheet of paper or card to protect the area from any spray which might overshoot. Spray evenly, from a distance of about 23 cm (9 in) until the required depth of colour is achieved. Always use aerosols with great care and follow the manufacturer's instructions for using them in confined spaces, disposing of the used cans, and any other special safety information.

With both the brush/sponge and spray techniques, colours can be mixed. This is especially effective over a large area, where they can be made to blend into one another. Shading can be created by using colour more strongly in some areas than in others. You can have a lot of fun experimenting with these kinds of effects.

Lift the stencil cleanly from the surface. It is ready to re-use almost immediately with modern, quick-drying paints.

STENCIL TEMPLATES
These designs can be used same size, or reduced on a photocopier. If you are using more than one colour, use acetate for the stencil and cut a stencil for each colour. The clear acetate will allow you to position each stencil precisely over what you have done already, for each colour application.

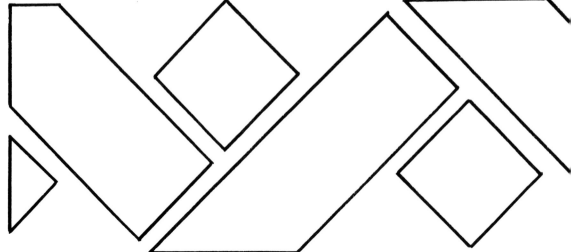

COOKERY CRAFTS

The countryside provides an abundance of wholesome and nutritious produce which can be used in a variety of different ways. Entertain your friends with refreshing home-made fruit drinks and wines. Capture the flavour of summer with jams, jellies and preserves. Fill the house with the delicious smell of home baking.

Fruit drinks and wine

Home-made fruit drinks are refreshing and good for you, and both non-alcoholic drinks and wines are an excellent way of using up a glut of fruit in the autumn.

For thousands of years people have made drinks, both alcoholic and non-alcoholic, from all kinds of fruit and other crops. The principle is always the same: to infuse water with the flavour of the fruit. It is then either drunk straight away, or fermented first. As long as you are sure that the fruit you are using is safe to consume, you can experiment with any kind of taste using the basic principles in these recipes.

Non-alcoholic fruit drinks are especially refreshing in summer, as well as being ideal for parties when it is good to be able to offer those who prefer not to drink alcohol a more exciting drink than the usual mixers and pops. These drinks are also very healthy, and are ideal for children, both at parties and for everyday consumption.

Wine-making is a fascinating hobby and many people experiment with it for years. Some wines have been developed over generations in the search for the most suitable fruit and the best water. Success is very much a matter of taste, and part of the fascination is in discovering which wines can go on being made with consistent results year after year. The ingredients can certainly be very cheap; and there is the added appeal of using hedgerow fruits, which make the vintner feel that here is 'something for nothing'.

NON-ALCOHOLIC PUNCH *

MAKES APPROXIMATELY 3 LITRES (5 PT).

Almost any fruit juice or non-alcoholic drink can be used to make up a punch. Experiment with different ingredients and quantities to find out what you like. The basic principle is always the same. Dissolve the sugar in the water with the spices to get the best out of their flavours, then allow the syrup to cool before adding it to the fruit ingredients. They taste better this way, and all the goodness in the fruit is retained. Finally, add any fizzy components, such as ginger beer, just before serving, so that they do not have a chance to go flat. Float a few interesting pieces of fruit on top to finish it off.

▲

300 ml ($^1/_2$ pt) pineapple juice
600 ml (1 pt) orange juice
juice and finely grated rind of 1 lemon
600 ml (1 pt) water
2.5 ml ($^1/_2$ tsp) mixed spice
2.5 ml ($^1/_2$ tsp) finely ground nutmeg
6 cloves
175 g (6 oz) sugar
1.1 litre (2 pt) ginger beer
1 orange

▼

Mix the pineapple and orange juice together in a large basin. Add the lemon rind and juice. Heat the water, adding the mixed spice, nutmeg, cloves and sugar, and stir until the sugar is dissolved. Allow to cool, then add to the fruit juice mixture. Place in a very cool place, or a fridge, to chill. Chill the ginger beer at the same time. When ready to serve, strain the fruit mixture into a large jug or punch bowl and add the ginger beer.

If Yorkshire people are given knives or scissors as gifts, they give a penny for each blade to the donor, so that they are paid for and will not cut the friendship.

Home-made fruit punch and lemon barley water are unbeatable as refreshing drinks in summer.

Add thinly sliced orange to float as decoration and serve with ice.

Use colours that mix well. For example, avoid mixing orange juice with blackcurrant juice. Although it tastes delicious, it looks rather unappetizing. All the citrus fruits look good together; and blackcurrant can be mixed with lemonade, ginger beer, or any red juice such as redcurrant.

Allow time to chill and a cooking time of approximately ¹/₂ hour.

LEMON BARLEY WATER *

MAKES APPROXIMATELY 1 LITRE (2 PT).
This is an absolutely delicious drink which should be served very cold in summer. It is much nicer than ordinary lemonade, so be sure to make plenty as it will soon be gone.

▲

25 g (1 oz) pearl barley
1.1 litre (2 pt) water
juice and rind of 1 lemon
50 g (2 oz) sugar (or to taste)

▼

Put the barley and water in a saucepan and bring to the boil. Simmer for an hour, then strain the water into a bowl or jug. Add the lemon juice and rind, and the sugar, and leave until cold. Strain to remove the rind.

Cooking time: approximately 1 hour.

Variation This recipe can also be used to make orange barley water by substituting an orange for the lemon. It is sweeter and more bland than the lemon, but still delicious.

WINES

Wines can be made from most kinds of fruit, flowers, leaves or roots provided that they are definitely non-poisonous. Always use freshly picked material and wash it well before using in case of contamination, especially if the material is from the roadside or farmland hedgerow. Many special ingredients are used in wine-making, but few are expensive, and most are available either from larger chemists, or from specialist wine shops. As

Delicious home-made wines are satisfying to make as well as being good to drink.

well as the ingredients, some special equipment, such as fermentation locks, will be needed and again these are easily found.

Never use any metallic equipment during the fermentation process, and ensure that all equipment is thoroughly sterilized before use to avoid the introduction of any kind of contaminating organism, such as an unwanted mould. To sterilize equipment, use either sodium metabisulphate or crushed Campden tablets, which can be bought from a chemist or specialist wine shop, and follow the instructions which accompany them carefully.

General recipe for wine

Using a plastic bucket or container for the initial fermentation, cut fruit or vegetables into small pieces and put them into this container. Cover with boiling water.

Add any additional flavourings at this stage; for example, raisins and/or thinly sliced citrus fruits if making a fruit wine. At this stage the peptic enzyme can be added if used.

Cover the container with a piece of clean cotton cloth (old sheeting is ideal), and tie it down firmly to keep out the Vinegar fly, a contaminating pest that will spoil the flavour of the wine.

Template for Emmerdale wine label

Stir the mixture, which is now called *must*, twice a day for three days, then strain it through a sheet of muslin or through a muslin bag into a clean, sterilized bowl.

Mix the yeast with a little sugar and water. Dissolve the rest of the sugar in a little water. When the yeast becomes active (when it begins to bubble and becomes creamy) add it to the strained liquid together with the sugar solution.

Take a fermentation jar. This is a glass jar with a special stopper which contains a fermentation lock – a curved glass tube which holds sterilizing water, through which the fermenting gases from the wine can bubble out, but which forms a barrier against any contaminating air from the outside coming into contact with the wine. Crush a Campden tablet into the water in the fermentation lock to act as the sterilizing agent. Stir the wine mixture well, and then strain it through a funnel into the fermentation jar.

Leave the wine in a place where it will not be disturbed, and will remain at an even temperature of 20°C (70°F) throughout fermentation. It should remain there until all fermentation has finished; that is, when there are no more bubbles coming through the fermentation lock.

Siphon the wine into another jar, leaving any sediment in the bottom of the original jar. This is usually done by sucking through rubber or plastic tubing to start the wine off along the tube, and then allowing it to continue to run from one vessel to the other.

Add two crushed Campden tablets to the wine and place the jar in a cold room for about three weeks, still with the fermentation lock in the stopper.

Replace the fermentation lock with a plain rubber bung and leave the wine to stand until it has cleared. If the cloudiness remains, buy isinglass fining gel and use it according to the instructions to clear the wine.

When the wine is clear, siphon it into bottles, again leaving any sediment in the jar. Cork the bottles, and label if wanted. Keep the wine for as long as possible, but certainly for at least six months, before drinking.

APPLE WINE***

MAKES APPROXIMATELY 2 LITRES (4 PT)

This wine can be made from windfall apples which would otherwise be thrown away. It is sweet and white, a real country wine. The sterilization of equipment is especially important with this wine, as it can be prone to spoiling by contamination.

▲

2.75 kg (6 lb) windfall apples
water
pectic enzyme
nutrients
wine yeast
1.5 kg (3 lb) white sugar
2 Campden tablets

▼

Trim off any bruised and damaged parts of the fruit. Wash and cut the fruit into small pieces. Place the fruit in a plastic container. Cover with boiling water, and leave to cool. When cool, stir in the pectin enzyme and continue stirring well. Cover and leave overnight.

Stir in the nutrients and the previously activated wine yeast. Leave the mixture for five days at a temperature of 20°C (70°F), stirring several times each day.

Strain off the liquid through muslin.

Stir in the sugar and bring the level up to 4.5 litres (1 gal) with water. Keep the mixture covered until the frothing ceases. Stir vigorously and pour into a fermentation vessel. Insert an air lock and hold at a temperature of 20°C (70°F) until the bubbles cease. Move the vessel into a cold room with the air lock still in place and leave for two to three weeks.

Siphon off the wine, leaving all sediment behind. Dissolve 225 g (½ lb) of sugar to each 4.5 litres (1 gal) of wine and add the crushed Campden tablets. The wine should be stored again with the air lock in place at fermentation temperature.

When the bubbles cease, sweeten to taste, replace the air lock and return to the warmth.

When fermentation ceases, transfer the wine to a cool place for two or three weeks before siphoning off.

Store the wine for about six months before bottling.

Variation This recipe can be used for many fruits and vegetables, especially the harder types which need the full flavour coaxed out in a long drawnout process.

Wine can be made from (almost) any surplus fruit or vegetables, so the next time you have a glut and have bottled, dried or frozen all you require, use the surplus to make a fine wine.

BLACKBERRY PORT WINE*

MAKES APPROXIMATELY 2 LITRES (4 PT).

Nothing is more typical of autumn than baskets of blackberries picked from the hedgerows. They have the added attraction that they are free, and do not require any cultivation. When you have made sufficient jams and jellies, try this easy recipe for an interesting taste. It is much more simple than the fruit wine, and it is interesting to compare blackberry wine made in this way with that made using the general recipe. This recipe is an old and easy farmhouse way of using up a blackberry glut.

▲

450 g (1 lb) blackberries
1.1 litre (1 quart) cold water
450 g (1 lb) sugar for each quart of liquid

▼

Place the fruit in a bowl and cover with water. Let it stand for 3 days, stirring 2 or 3 times each day.

Strain off the liquid and then press the remaining mixture through a straining bag.

To each 1.1 litre (quart) of liquid add 450 g (1 lb) of sugar and let the mixture stand for another 3 days, stirring as before.

The liquid can now be bottled.

Variation Try this very simple recipe with other soft fruits that easily yield their juices; blackcurrants and raspberries are good examples. These fruits will make drinks that preserve the pure taste of the fruits.

Jams and Jellies

Whether you grow your own produce or take advantage of cheaper prices in season, mastery of the traditional country skills of jam-making will allow you to enjoy the fruits of summer all year round.

Jams and jellies have always been as much a part of the northern ritual of tea as baking. They provide an easy way of preserving fruit, and are much more delicious than commercially produced jams. If you are able to use fruit from your own garden they are cheap, and are a good way of using up surpluses. Some of these jams are so delicious, however, they are worth making with frozen fruit out of season. Unless you have a very exotic garden you will need to buy lemons of course, but they are used so much in country cooking because they are useful, and delicious.

Jam-making is one of the easiest of all country activities and even those who swear that they cannot cook will find jam-making satisfying; and, more important, will enjoy the results.

Older Yorkshire folk call the large, yellow root vegetable, which is so delicious mashed with butter and pepper, a turnip. This causes confusion down South where it is usually known as a swede.

Jam-making is an excellent way to use up surplus fruit, and requires no special equipment.

To cover and seal the newly made jam you can either use existing metal screw tops on the jars, or buy paper rounds and cellophane discs which are held on with an elastic band. Either way, the jars will look most attractive if the top is covered with a circle of gingham or other fabric tied on with fine ribbon.

Jam should always be boiled in a saucepan large enough so that the mixture does not come more than half-way up the sides, to allow a full rolling boil without the jam coming over the top. If you are planning to make a large quantity of jam, it is worth investing in a preserving pan, one of the large, old-fashioned pans, which used to be made of brass but are now copper or aluminium. They usually have a handle over the top rather than at the side to make them easier to lift. Do remember that jam is extremely hot when it is cooking and should always be treated with respect.

Setting point

Jams and jellies always consist of a mixture of fruit and sugar which is vigorously boiled until it is ready to set when cooled. There are various ways of deciding whether or not this 'setting point' is reached. Experienced jam-makers can tell by the noise. The 'bubble, bubble' noise of jam at a rolling boil changes to a 'gloop, gloop' when the jam is ready. However, this is not a very reliable test for beginners, and there are more scientific ways. Many people put a little jam on a cold saucer. If, after about a minute, the jam begins to form a skin strong enough to wrinkle when pushed with the finger, it is ready. Another good test is to lift some jam on a wooden spoon and pour it back into the pan. At the start, before the jam is ready, it will simply pour like any other liquid; but as setting point is reached, it will gather in large jelly-like drops and fall reluctantly.

Setting point becomes easier to assess the more jam you make. It does not matter too much if you get it wrong to start with. Your equally delicious jam will just be rather difficult to keep on the buttered toast. If you make a batch of very runny jam, you can return it to the pan and boil it again to achieve a better result.

RHUBARB JAM WITH APRICOTS*

MAKES APPROXIMATELY 1.75 KG (4 LB).

One of my abiding memories of the Yorkshire landscape is of acres and acres of rhubarb, stretching in great lines over the fields, and being forced for the early market in the dark of the old air-raid shelters. Rhubarb is very happy in many Dales gardens, so it had to feature in at least one of the jams.

▲

900 g (2 lb) fresh rhubarb, wiped and chopped
900 g (2 lb) sugar
225 g (¹/₂ lb) dried apricots
150 ml (¹/₄ pt) water

▼

Layer the rhubarb in a basin with the sugar. Put the dried apricots in another basin and cover with 150 ml (¹/₄ pt) of boiling water. Leave both of these mixtures for 24 hours.

Put the rhubarb and sugar mixture and the soaked apricots into a preserving pan and bring to the boil, stirring until all the sugar is dissolved. Keep at a full rolling boil for about ³/₄ hour, or until

On Shrove Tuesday morning,
from 11.00 to 11.05,
Bingley Parish Church bell
will ring to signal to
housewives that it is time
to begin to make
their batter.

Delicious home-made jams and jellies will not last on the shelf for long.

setting point is reached (see page 60). Pour carefully into warmed jars and cover immediately.

If you like this jam with ginger, add about 40 g (1½ oz) of finely chopped fresh or crystallized ginger before cooking.

Allow about 24 hours, which includes a cooking time of ³/₄ hour.

REDCURRANT JELLY*

MAKES APPROXIMATELY 900 G (2 LB).

Our children eat this on bread and butter, but it can also be used as a delicious accompaniment to meats such as lamb or game. Jellies are a very palatable alternative to jams, but they are not as economical because only the juice is used. However, the fruit generally requires less preparation. It is better to pot jellies up into smaller jars because they become runny as they are used.

▲

675 g (1½ lb) redcurrants, topped and tailed
300 ml (½ pt) water
sugar to be weighed later

▼

Put the redcurrants in a preserving pan with the water. Bring to the boil and simmer until the fruit is cooked and soft. Pour into a jelly bag suspended over a basin and leave until the following day (see page 18 for notes on using jelly bags). Do not squeeze the fruit pulp in the bag as this makes the jelly cloudy. Measure the juice which has run out of the fruit, and then replace it in the preserving pan. Add 450 g (1 lb) of sugar to the juice for each 600 ml (1 pt) of juice. Bring to the boil, stirring until the sugar is dissolved, and keep at a full rolling boil until setting point is reached. Pour carefully into warmed small jars and cover immediately.

If you have no jelly bag, you can use a sheet of strong gauze or any strong, natural-fibre, finely woven white fabric. Place the jelly bag or fabric in the bowl, which must be large, and then pour the fruit mixture into it. If you have no convenient place from which to hang the bag, then an upturned stool is very good, hanging the fabric from the legs. Be sure to tie the bag or fabric up very firmly as it makes a mess if it falls into the juice.

Allow 24 hours, which includes a cooking time of about ³/₄ hour.

RASPBERRY PRESERVE*

MAKES APPROXIMATELY 2 KG (4½ LB).

Raspberry bushes love the cold winters and sunny summer slopes of the Dales, and many happy childhood hours have been spent picking the fruit, not to mention eating it. Raspberries are so much nicer than the overrated, rather bland strawberry, and they are wonderful just on their own or with cream. Raspberry jam, though, is *the* jam, featuring in all the songs and stories, and this version is the most delicious of all.

▲

900 g (2 lb) raspberries
1.25 kg (2½ lb) sugar

▼

Put the raspberries in a large ovenproof bowl, and put the sugar into a second ovenproof bowl. Place both in a moderate oven until the fruit and sugar are very hot. Remove the fruit and, while still hot, beat thoroughly. Continuing to beat

Template for Jam Label

throughout, add the hot sugar gradually until it is all mixed in. Pour carefully into warmed jars and cover immediately.

This preserve is extremely easy to make, and retains all the wonderful flavour of the fruit because it is not cooked. It does not keep as well as a true jam – but it is unlikely to get the chance as it is so delicious.

Allow about ³/₄ hour, which does not include any cooking time.

LEMON CURD**

MAKES APPROXIMATELY 1.5 KG (3 LB).
Sometimes also known as lemon cheese, this is a delicious mixture of lemon, egg and sugar cooked very gently into a smooth thick cream. When you make your own it will have a rich, sunshine yellow colour rather than the sharp acid yellow of the kind on the supermarket shelves. It tastes absolutely wonderful, and keeps fairly well.

▲

finely grated rind of 6 lemons
sieved juice of 4 lemons
100 g (4 oz) butter
900 g (2 lb) sugar
6 eggs

▼

Put all the ingredients except the eggs into the top half of a double-boiler, with water in the bottom half. If you do not own a double-boiler, use a heat-proof basin which will fit into a large saucepan without touching the bottom of it. Heat, stirring until the sugar is dissolved. Beat the eggs and stir in gradually. Cook gently for an hour, stirring occasionally, during which time the mixture will gradually thicken. Pour the mixture carefully through a sieve into warmed jars and cover immediately.

Cooking time: about 1¹/₂ hours.

Variation Lemon curd is the most familiar of these recipes, but curd can be made in the same way using oranges, or a mixture of oranges and lemons.

Pig killing was always a time of great celebration as an excuse for the feasting which resulted from the using of all the bits of the pig which could not be cured. Nothing is wasted on a pig. Everything that could be kept was, of course, but perishable items like the blood had to be used immediately and blood was usually made into black pudding, a sausage made in the gut with the blood, flour and cubes of fat. Pig killing time was very cooperative and items would be shared and exchanged, with one wife making black pudding, another pork pies and so on, and these were often gifts and exchanges. It is very bad luck to return a basin from such a gift washed. It should always be returned dirty.

Chutneys and Preserves

Delicious chutneys and preserves are not difficult to prepare. They help to use up the avalanche of autumn fruit and provide a delicious accompaniment to all kinds of dishes from cold meat to curries.

These recipes are simple to make, and will remind you throughout the long winter of the warmth of summer. They are as easy to make and as rewarding as jams, and add a special home-made touch to all kinds of dishes from the simple salad and cold meats to the more exotic curry.

TOMATO KETCHUP***

MAKES APPROXIMATELY 1.1 LITRE (2 PT).

All along the East Yorkshire coast are the seaside resorts where both Yorkshire people and 'foreigners' have enjoyed annual holidays and day trips for most of this century. And while there, they have sampled the Yorkshire feast of fish and chips, bread and butter and a pot of tea, the highlight of which is home-made tomato ketchup. Home-grown tomatoes taste super in this ketchup and give it a distinctive colouring. It is a much softer colour than commercially produced ketchup, and goes equally well with hot or cold food.

▲

1.5 kg (3 lb) ripe tomatoes, chopped
450 g (1 lb) cooking onions, chopped
175 g (6 oz) white sugar
3 tbsp mustard powder
3 cloves of garlic, crushed
1 tsp salt
150 ml (¹/₄ pt) red wine vinegar

▼

Mix all the ingredients in a large saucepan. Bring the mixture to the boil, and simmer uncovered for 45 minutes, stirring occasionally. Remove from the heat and allow to cool so that it is easier to handle. Pour the mixture into a liquidizer and blend into a purée. Press the purée through a sieve and return it to the washed-out saucepan. (If a liquidizer is not available, the mixture can be pressed through the sieve twice.) Bring to the boil once more stirring occasionally, then remove the saucepan from the heat and allow to cool again. Pour the ketchup into warmed bottles or jars and cover immediately.

APPLE CHUTNEY***

MAKES APPROXIMATELY 1.75 KG (4 LB).

Apples are a very versatile fruit that can be dried, bottled or used to add bulk and sharpness to jams and jellies. This delicious apple chutney brings out the taste of the fruit and is a piquant addition to any hot or cold meat or cheese dish.

▲

900 g (2 lb) of apples, peeled, cored and diced
450 g (1 lb) demerara sugar
600 ml (1¹/₂ pt) vinegar
350 g (³/₄ lb) seedless raisins
25 g (1 oz) mustard seed
15 g (¹/₂ oz) garlic, peeled and finely chopped
2 tsp salt
2 tsp ground ginger
2 tsp cayenne

▼

Put all of the ingredients into a preserving pan and boil for two hours, stirring from time to time to prevent burning. Remove the pan from the heat, pour the mixture into warmed jars and cover immediately.

Variation This recipe is best made with cooking apples, and you can vary the taste by using different varieties of vinegar. Instead of malt vinegar, you could try clear malt vinegar, a wine vinegar or a cider vinegar.

PICKLED ONIONS*

With all of the super cheeses that Yorkshire is famous for, what better accompaniment for a beer and cheese lunch or supper than small, home-pickled onions prepared in your own kitchen and

Chutneys and preserves are simple to prepare and complement good country cooking.

perhaps even grown in your own garden. The smaller varieties of onions have the best flavours for pickling.

▲

*small, fresh silverskin onions (as many as you
require)*
salt
white wine vinegar
white peppercorns
stick of ginger (optional)

▼

Soak the onions in cold salted water for a few hours to make them easier to peel. Remove layers of peel until the onions look clear, taking care not to cut the bulb.

Add a teaspoon of white peppercorns to each 600 ml (1 pt) of vinegar being used. Pack the onions into jars and cover with the vinegar. An optional 2.5 cm (1 in) stick of ginger may be

Mince pies should have an oval pastry shell, to represent the shape of The Manger, with the richness of the mincemeat inside symbolizing the rich gift of Jesus. Traditionally, one should eat one mince pie on each of the twelve days of Christmas, bringing good luck for each of the twelve months to come.

Home-made mincemeat makes extra-special mince pies. Make the mincemeat several weeks in advance of Christmas for the best results.

added to each 600 ml (1 pt) jar of pickles. Put lids that will not be affected by the vinegar on the jars. Store for three days and then pour off the vinegar and boil it. Refill the jars, covering the onions with the boiled vinegar. Allow to cool, and lid the jars again. Store for a few weeks before using.

As a finishing touch, you can add a sprig of tarragon to each jar, which will provide a nice colour contrast.

MINCEMEAT*

MAKES ABOUT 2.25 KG (5 LB) OF MINCEMEAT.
No Christmas would be complete without mince pies, and the delicious filling for these is usually made in the autumn, when apples are plentiful. The rich taste of this delicious mixture has no equal, and it is very easy to make. It gets its name from the fact that at one time it really did contain meat, but this is not attractive to modern tastes and now the only meat ingredient is the suet. This is usually beef suet, but vegetarian suet (easily obtainable from health food shops and some supermarkets), can be substituted without affecting either taste, appearance or texture.

▲

*sufficient lemons to yield 100 g (4 oz) of grated
peel (between 6 and 8 depending upon size)*
450 g (1 lb) apples, peeled and finely chopped
450 g (1 lb) currants
1 nutmeg, finely grated
225 g (8 oz) sultanas
225 g (8 oz) raisins
450 g (1 lb) sugar
125 g (4 oz) suet

▼

Place the lemon peel and chopped apples in a large bowl and mix well. Add the rest of the ingredients and mix thoroughly. Allow the mixture to stand for two hours, then pack into jars and cover with lids.

Mincemeat made in the autumn is usually kept until Christmas of the same year, but it will last longer if it is unopened. It should be kept for several weeks before eating, to allow all the flavours to blend together.

BOTTLED DESSERT FRUIT **

This is a lovely way to preserve fruit to provide a wonderful winter sweet.

▲

1 bottle of white rum or brandy
granulated sugar
fruit of your choice – You may use a selection of fruit, but it is best to avoid citrus fruits, apples, bananas and pears. It is also possible to layer some fruit in the jar, cover with spirit, and wait for another fruit to come into season before adding that and repeating the process. Use a large stone jar or, better still, a large, glass preserving jar.

▼

Weigh an empty jar.

Carefully choose the best ripe fruits. Do not peel or stone the fruit but wipe gently. Put the fruit into the jar in attractive colour layers.

Weigh the full jar, and subtract from this the weight of the empty jar to find the weight of the fruit. To the brandy or rum in a jug add sugar equivalent in weight to half the weight of the fruit. Dissolve the sugar in the liquor (this can be speeded up by placing the jug in a bowl of hot water). Cover the fruit in the jar with the sugar liquid; cover and store in a cool place. Other fruits may be added as they come into season. The fruits are best kept for at least three months.

Bottled fruit is a good way to use up a glut. Bottling also captures the taste of summer so that it can be enjoyed all the year round.

Baking

Evoke memories of the past, and satisfy hearty country appetites, with wonderful home-baked cakes, breads, and other traditional recipes.

Baking has always been a mainstay of Yorkshire cooking, whether it is for tea-time treats, beautiful breads which are always so much more delicious than those bought in a shop, or a baked pudding to go with a meat course. Baking produces an evocative smell and there is nothing to beat it on a cold day when people are hungry.

YORKSHIRE PUDDING * *

MAKES APPROXIMATELY 12 SERVINGS.

Probably the most famous Yorkshire dish of all, Yorkshire Pudding is not really a pudding at all, of course. Now it is usually eaten as an accompaniment to roast beef with all the trimmings, but traditionally it is eaten separately as a first course, either with gravy, or with a sweet sauce like Raspberry Vinegar (see page 18). Yorkshire Pudding can be made in individual bun trays (in which case, reduce the cooking time a little). A true Yorkshire Pudding, however, is square and huge, heaped with great crusty bubbles and served piping hot.

▲

75 g (3 oz) plain flour
¹/₂ tsp salt
2 eggs
300 ml (¹/₂ pt) milk
15 g (¹/₂ oz) lard or dripping to cook pudding
shallow baking tin
(An alternative ready measure is 1 teacup of plain flour and 1 teacup of milk to each egg, plus salt.)

▼

Pre-heat oven to 220°C/425°F/gas 7.

Sieve flour and salt into a basin, and make a well in the centre. Drop the eggs into the well. Mix in the eggs carefully, drawing the flour into the eggs gradually and adding milk as necessary. The mixture should form a smooth, lump-free thin paste when all the milk is finally added.

Beat well until the mixture is smooth and full of fine air bubbles.

Heat the dripping or lard in the baking tin until extremely hot. Pour in the pudding mixture and place in the top of the oven immediately.

Do not open the oven door for 20 minutes. Remove when well risen and golden brown.

Cooking time: approximately 25 minutes.

Variation Various herbs can be added to the pudding batter before cooking to make a savoury version; for example, chopped and cooked onions with sage.

SULTANA SCONES *

MAKES APPROXIMATELY 20 SCONES.

There are many ways of making scones, and they can be flavoured with anything from ham to treacle, but this is a good basic recipe with sultanas for this tea-time treat. If the sultanas are omitted a

There was a superstition in Southern Yorkshire that Yorkshire Pudding should always be mixed and beaten on the doorstep.

Yorkshire Pudding – roast beef is not complete without it. Traditionally, it was eaten on its own with gravy or a sauce such as Raspberry Vinegar.

good plain scone is made. Scones are delicious simply buttered, especially while still hot, but can also be eaten with jam and/or cream cheese, or with honey, or with almost anything else.

▲

450 g (8 oz) white self-raising flour
¹/₂ level tsp salt
50 g (2 oz) butter
25 g (1 oz) sugar
50 g (2 oz) sultanas
150 ml (¹/₄ pt) milk
extra milk for brushing

▼

Pre-heat the oven to 230°C/450°F/gas 8.

Sift the flour and salt into a bowl and rub in the butter to produce a fine crumbly mix. Mix in the sugar and sultanas.

Add the milk all at once and mix with a knife into a soft dough.

Turn out onto a floured board and knead lightly until smooth, then roll out to about 1 cm (¹/₂ in) in thickness.

No Yorkshire people call small bite-sized buns 'cakes'. A cake is large enough to slice, and only foreigners call individual tea-time treats cakes. They are buns. Really elaborate iced ones might be referred to as 'fancies'.

Cut into rounds with a cutter and place on a greased baking tray.

Brush tops with milk and bake for about 10 minutes, until the scones are golden brown and well risen.

Cool on a wire rack.

Cooking time: about 10 minutes.

Note: If preferred, the scone mixture can be cooked in large rounds rather than cut into smaller ones. To do this, shape the dough into two or three flat rounds about 1 cm (¹/₂ in) thick and score the top to mark it into triangular pieces before cooking.

YORKSHIRE TEACAKES * *

MAKES 6

When I was a child in Yorkshire one of the greatest treats, when out for the day, was to be taken into the local tea-rooms for a large cup of strong tea, and a very hot teacake, split, toasted and richly buttered.

▲

25 g (1 oz) fresh yeast, or the equivalent of dried
yeast
1 tsp sugar to mix yeast
450 g (1 lb) plain flour
1 tsp salt
25 g (1 oz) lard
25 g (1 oz) sultanas
25 g (1 oz) currants
25 g (1 oz) sugar
300 ml (¹/₂ pt) milk
extra milk to brush

▼

Pre-heat the oven to 220°C/425°F/gas 7.

Cream the fresh yeast with the teaspoonful of sugar or, if using dried yeast, prepare as directed using part of the milk. Sieve the flour and salt into a mixing bowl. Rub in the lard. Add the sultanas, currants and sugar and lightly toss together. Warm the milk (it must not be hot) and add gradually to the creamed yeast. Make a well in the dried ingredients and add all of the liquid. Mix into a smooth dough. Cover the basin with a cloth

An irresistable tea-time spread of scones and hot, buttered teacakes with raspberry jam.

and leave in a warm place for about an hour, or until the dough has risen to fill the basin.

Turn out onto a floured surface and knead well. Divide into six equal pieces and knead each into an even, round piece. Roll each piece out into a 15 cm (6 in) round and place, well spaced, onto a greased baking sheet. Brush tops with milk. Leave in a warm place until doubled in size. Bake in the centre of the oven for about 15–20 minutes, until golden brown.

Yorkshire teacakes really should have dried fruit in them, but you can, of course, omit it if you prefer.

Final cooking time: approximately 15 minutes.

RICH FRUIT CAKE**

This is a very old recipe with an unusual method of cooking which gives a delicious rich taste. It is served in Yorkshire just as it is, but always with cheese – Wensleydale of course.

▲

165 g (5½ oz) margarine
450 g (1 lb) mixed dried fruit
165 g (5½ oz) demerara sugar
1 level tsp mixed spice
1 level tsp powdered cinnamon
300 g (11 oz) self-raising flour
4 eggs
³/₄ wine glass water (or white wine if preferred)
175 g (6 oz) glacé cherries, halved and coated in
a little of the flour
20 cm (8 in) diameter round cake tin
greaseproof paper
cooking foil

▼

Pre-heat the oven to 170°C/330°F/gas 3.

Prepare the tin by lining it with greased greaseproof paper. In a large saucepan heat the margarine, dried fruit and demerara sugar. Bring to a gentle simmer, stirring all the time, then add the mixed spice and cinnamon. Simmer for 30 minutes, stirring occasionally. Remove from the heat and allow to cool until warm rather than hot. Sieve flour into a basin. Beat eggs and water or

wine together in a separate basin. Add small quantities of flour and then egg mixture alternately to the fruit and sugar mixture, stirring well, until the flour and eggs are used up. Fold the cherries gently into the mixture. Pour into the lined cake tin. Cover the tin with a piece of foil to prevent the cake drying out, and bake in the centre of the oven for 1 hour. Remove the foil and cook for approximately 1 further hour, or until a steel knitting needle or sharp knife pushed into the centre of the cake comes out dry and clean.

Although this kind of cake is served in the Dales with nothing but cheese to accompany it, it can of course be marzipanned and/or iced. It also keeps very well either in an airtight tin, or in a freezer.

Final cooking time: approximately 2 hours.

PIKELETS**

MAKES APPROXIMATELY 12 PIKELETS.

There is much confusion surrounding the difference between crumpets and pikelets, but in fact there is very little difference between them. Both are small, round cakes full of air holes. They are delicious eaten dripping with butter whilst still

The first real York Ham was cured in the smoke from burning the sawdust made in the cutting of the great timbers for York Minster.

This rich fruit cake is traditionally eaten with Wensleydale cheese.

hot, or, if eaten later, they can be toasted first. Butter is probably enough, but they can also be eaten with jam, honey or marmalade. They are great fun to make, being cooked in a special ring which contains the batter and inside which the batter rises and bubbles spectacularly. The technique is very simple but it may take some practice to get the appearance right, though the results will always be delicious.

▲

7 g (¹/₄ oz) fresh yeast, or the equivalent of dried
yeast
¹/₂ tsp sugar
300 ml (¹/₂ pt) warm (not hot) water
275 g (10 oz) plain white flour
15 g (¹/₂ oz) salt

▼

You will need some pikelet or crumpet rings, metal rings which sit on the base of a frying pan and contain the batter. They are similar to the ones used by some cooks to contain frying eggs.

Cream the fresh yeast with the sugar and a little of the warm water, or make up the dried yeast as directed.

Sift the flour and salt into a large basin, mix well and make a well in the centre. Add about half of the remaining water to the yeast mixture and mix well, then add this liquid gradually to the dry ingredients, pouring it into the well, and mixing the flour in slowly to form a lump-free batter. Add the remainder of the water when the yeast mixture is all used up, making a thick batter.

Cover with a cloth and leave to rise until doubled in volume.

Grease the crumpet rings and place in a hot greased frying pan, which must have a very flat base to prevent the batter escaping from the bottom edge of the rings.

Pour batter into each ring to about half way up the side, and cook until dry on top.

Carefully remove the rings and turn the pikelets over until both sides are evenly brown and cooked.

Cooking time: approximately 30 minutes, including rising.

SCOTCH PANCAKES *

MAKES APPROXIMATELY 12 PANCAKES.

These are called by all sorts of names in different areas, but are usually known as Scotch pancakes in the north of England. They are similar to ordinary pancakes, but are much smaller and thicker, and are meant to be eaten, warm or cold, with butter.

▲

100 g (4 oz) white self-raising flour
1 pinch of salt
2 tbsp sugar
1 egg
30 ml (2 tbsp) milk (extra milk may be needed
to mix)

▼

Mix all the dry ingredients in a basin and make a well in the centre. Beat the milk with the egg and add to the well in the dry ingredients. Mix in gradually, adding more milk if necessary until the mixture is a soft dropping consistency.

Grease a frying pan well and allow it to become very hot. Pour enough of the mixture into the pan to form a small pancake about 10–13 cm (4–5 in) across. Turn after a few moments to cook the other side.

Cool on a wire rack.

Cooking time: approximately 10 minutes.

"Tha mun nivver spoil
a good bit o' Wensleydale
wi' cookin' it."

Pikelets (left) and Scotch pancakes (right) hot off the griddle. Eat with lashings of butter for a special treat.

TRADITIONAL LARDY CAKE * *

This cake is gloriously sticky, extremely bad for you, and delicious. In the days before people worried about cholesterol, lardy cake was an excellent way to provide a large quantity of warming and filling food for a hungry family fairly cheaply. Now, this poor man's staple has become everyone's occasional guilty indulgence.

▲

white bread dough made in the same way as given on page 82 using 450 g (1 lb) of flour and risen until doubled in size for the first time
175 g (6 oz) lard, chopped into small pieces (or hardened in the freezer and grated)
175 g (6 oz) mixed dried fruit
50 g (2 oz) candied peel
175 g (6 oz) sugar
large rectangular baking tin

▼

Pre-heat oven to 220°C/425°F/gas 7.

Turn the dough out onto a floured board and roll out with a rolling pin into a large rectangle. Spread two-thirds of this rectangle of dough with one-third of each of the other ingredients. Fold the uncovered third of the dough onto the covered dough, and then fold over again to produce a triple sandwich of dough. Press the ends together firmly to seal. Roll out this dough sandwich, and repeat the process with another third of each of the other ingredients over two-thirds of the dough. Fold in the same way as before, seal the ends, and repeat the process once more.

Place on a large, greased rectangular baking tray and leave in a warm place, covered, to rise for about 30 minutes. Then bake for about 45 minutes.

Remove from the oven and leave to stand for about 10 minutes before cutting into squares. Traditionally it is eaten warm, but is also decadently delicious cold.

Cooking time: approximately 45 minutes.

YORKSHIRE PARKIN *

This is a wonderful sticky tray cake, served cut into squares, which for us was part of the fun on Bonfire Night. But who needs an excuse? It is delicious all the year round. The flavour is improved if it is kept in a tin for a week before eating, but in practice this never seems to happen.

▲

100 g (4 oz) butter
100 g (4 oz) soft brown sugar
100 g (4 oz) treacle
100 g (4 oz) golden syrup
1 egg
150 ml (¹/₄ pt) milk
225 g (8 oz) plain flour
¹/₂ level tsp ground ginger
¹/₂ level tsp mixed spice
225 g (8 oz) oatmeal
¹/₂ tsp bicarbonate of soda
rectangular baking tin (about 28 × 18 cm/ 11 × 7 in)

▼

Pre-heat oven to 180°C/350°F/gas 4.

Put the butter, sugar, treacle and golden syrup in a pan and melt over a low heat until the sugar is just dissolved. Allow to cool slightly. Beat the egg well with half of the milk and stir into the syrup mixture. Sieve the flour, ginger and mixed spice into a mixing bowl and stir in the oatmeal. Pour the syrup mixture into the dry ingredients. Dissolve the bicarbonate of soda into the remainder of the milk and add to the rest of the mixture. Mix together well. Pour into the greased baking tin. Bake for about 1 hour in the centre of the oven, or until firm. Allow the parkin to cool before removing it from the tin. When cold, cut into squares.

Variation This recipe makes a rich, palatable parkin, but if you prefer a really dark and strong mixture you could use 225 g (8 oz) of treacle instead of the treacle and golden syrup mixture. Alternatively, if you use 225 g (8 oz) of golden syrup, you will make a lighter cake.

Cooking time: approximately 1 hour.

Lardy cake and Yorkshire parkin (top), two sticky and delicious traditional country recipes.

WHITE BREAD**

MAKES 1 450 G (1 LB) LOAF.

This basic loaf is wonderful just buttered, but can of course be eaten with any of the usual accompaniments.

▲

15 g (¹/₂ oz) fresh yeast or the equivalent in dried yeast
300 ml (¹/₂ pt) warm (not hot) water
1 level tsp sugar
450 g (1 lb) plain white strong bread flour
2 level tsp salt
15 g (¹/₂ oz) butter
a little milk for brushing
450 g (1 lb) loaf tin

▼

Pre-heat the oven to 230°C/450°F/gas 8.

Mix the fresh yeast to a smooth creamy paste with a little of the warm water and the sugar; or prepare the dried yeast according to the instructions.

Sift the flour and salt into a bowl, and rub in the butter until it has all disappeared into a fine crumb mixture.

When the yeast has bubbled into a creamy paste, blend in the rest of the water. Make a well in the centre of the flour mixture and add all the liquid at once. Mix together into a firm dough which should leave the sides of the bowl clean. If the dough is sticky, sprinkle in a little more flour.

Turn out onto a floured board and knead hard for ten minutes.

Return to the bowl and cover, leaving in a warm, but not hot, place to rise until the dough doubles in size.

Turn out and knead again until the dough feels smooth and firm. Shape into a loaf shape and place in the greased loaf tin.

Cover and leave to prove, or rise, again until the loaf fills the tin.

Brush gently with milk and place in the centre of the oven.

Bake for about 40 minutes until the loaf will come cleanly out of the tin and has a golden crust. A cooked loaf, if gently tapped on the base, sounds hollow.

Allow to cool on a wire rack.

Cooking time: approximately 1 hour.

WHOLEMEAL BREAD**

This is made in exactly the same way as white bread, substituting wholemeal flour for strong white bread flour.

Neither white nor wholemeal bread needs to be baked in tins. After the second kneading and before the second proving, the dough can be shaped in any way. Traditional shapes include a cottage loaf and a plait. To make a cottage loaf, shape two-thirds of the dough into a ball, and the remaining third into another ball. Place the smaller ball on top of the larger and push the long handle of a wooden spoon vertically down the centre of both balls to join them and to give the distinctive dimple in the middle of the top. To make a plait, divide the dough into three and roll each piece into a long thin sausage. Firmly join them together at one end, moistening very lightly with water if necessary, then plait the rolls and join the three ends at the other end. Cottage loaves, plaits and any other shapes can then be baked as described above, but they will need a slightly shorter baking time if the loaf is thin.

It is tempting fate to be so profligate as to cut a new loaf before the old one is all gone.

Nothing beats the mouth-watering aroma, or the flavour, of fresh-baked home-made bread.

NEEDLE CRAFTS

There is no prettier way to produce decorative and useful items around the home than with embroidery and crochet, patchwork and knitting. Sew a patchwork cot quilt for a new arrival. Knit up unusual cushion covers in traditional Aran patterns. Or create pretty tapestry bags and pictures.

Patchwork

More than any other craft, patchwork combines practicality with attractiveness, making use of treasured scraps of fabric to create new items to cherish, from cot quilts to cushion covers.

This absorbing craft was begun by country-women for whom materials were expensive and heated bedrooms unknown. Patch-work used small amounts of fabric, often the best scraps from discarded garments or household linen, which were sewn together to make a new bedcover or quilt. Because it is so much more satisfying to make something beautiful, wonderful patterns developed, such as the traditional Mosaic and Log Cabin used here. Sometimes, squares were simply arranged in attractive colour patterns, as on the modern cushion (page 88).

A lovely idea for patchwork is to create a quilt over many years, adding pieces from treasures such as wedding and christening gowns, first dresses and favourite shirts. Not all fabrics are suitable, however, especially when mixed together. Delicate fabrics need to be used with one another, and tough fabrics are also better together. Depending on the use to which the patchwork is to be put, very delicate fabrics should be avoided altogether, as should loosely woven fabrics such as very soft, coarse tweeds, which will disintegrate when cut into small pieces. Very thick fabrics should not be used as the seams will be too bulky for the size of the pieces. The best fabrics are dress cottons, fine fur-nishing cottons and fabrics of similar weight.

Complete patterns are given here that you can adapt and experiment with as you wish and as your material allows. Quantities are given for the size and pattern used here, and you can adapt them as you wish.

Special notes

Patterns which use only straight lines have been chosen because they are more suitable for a begin-ner. If you become hooked on this beautiful craft,

there are many traditional patterns using shell shapes, curves and diamonds, or you can make up your own.

Make sure you cut the pieces very accurately so that they fit together perfectly. The best way to do this is to use a template – a piece of card cut to the size and shape of the fabric pieces needed for the pattern. You then draw round it onto the fabric, taking care not to pull it out of shape. Cut care-fully along the drawn lines using sharp scissors, or a special patchwork cutter if you can get one. Place the template so that you use the fabric economically.

When making the template, add on a seam allowance.

You may feel a template is unnecessary for square or rectangular pieces. In this case, it is helpful to draw a thread from the fabric to estab-lish a straight line along which to cut your rec-tangles or squares.

The patchwork projects were assembled using a sewing-machine, but you may prefer to stitch the pieces together by hand. If so, a fine backstitch is

Patchwork cushions can be bright and modern (left) or traditional (right).

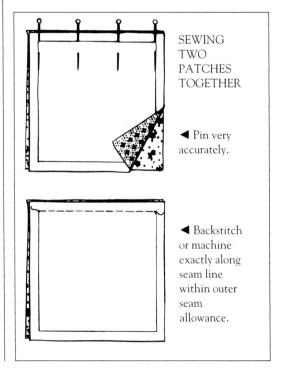

SEWING
TWO
PATCHES
TOGETHER

◀ Pin very accurately.

◀ Backstitch or machine exactly along seam line within outer seam allowance.

probably the best stitch to use. When assembling, place the pieces of fabric right sides together and sew very accurately along the seam allowance (see diagrams, p. 86), taking great care that the pieces are correctly positioned both in relation to each other, and also in relation to the design of the whole quilt. Very accurate placing and sewing of every piece makes all the difference to the appearance of the finished quilt.

Take care that all seaming is worked very accurately so that the final patchwork lies flat.

The seams will be stronger if they are pressed over to one side rather than open. Choose a pattern for pressing and use it throughout to give the quilt a uniform appearance. For example, on a cushion cover made of squares, press all the seams in the same direction across the cushion throughout.

LAYOUT FOR
MODERN CUSHION

PN	SF	LF	PP	LF	SF	PN	SF	LF	PP
SF	LF	PP	LF	SF	PN	SF	LF	PP	LF
LF	PP	LF	SF	PN	SF	LF	PP	LF	SF
PP	LF	SF	PN	SF	LF	PP	LF	SF	PN
LF	SF	PN	SF	LF	PP	LF	SF	PN	SF
SF	PN	SF	LF	PP	LF	SF	PN	SF	LF
PN	SF	LF	PP	LF	SF	PN	SF	LF	PP
SF	LF	PP	LF	SF	PN	SF	LF	PP	LF
LF	PP	LF	SF	PN	SF	LF	PP	LF	SF
PP	LF	SF	PN	SF	LF	PP	LF	SF	PN

MODERN PATCHWORK CUSHION***

This arrangement of joined squares is probably the most simple patchwork pattern of all. The squares can be of any size and colour arrangement, but for a modern effect diagonal lines of colour are attractive. Squares are also very effective if the fabric has a motif, and one motif can be placed in each square. The edges of this cushion were trimmed with piping made from cord and bias-cut strips of one of the fabrics. This is not essential, but it does give a neat, professional finish. If preferred, one side of the cushion cover could be closed with a zip or other fastening for removal.

Fabric code
PN = plain navy 1 piece 20 × 20 cm (8 × 8 in) or equivalent
PP = plain pink 1 piece 20 × 25 cm (8 × 10 in) or equivalent
LF = large floral pattern 1 piece 30 × 30 cm (12 × 12 in) or equivalent
SF = small floral pattern 1 piece 30 × 30 cm (12 × 12 in) or equivalent
Plus: of chosen backing fabric, 1 piece 40 cm (16 in) square; of fabric chosen for piping sufficient to cut a bias strip 3.5 × 152.5 cm (1½ × 61 in).
Approximately 152.5 cm (61 in) piping cord (piping is optional).

TEMPLATE FOR MODERN CUSHION (actual size)

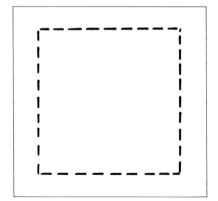

Cushion pad approx. 38 cm (15 in) square, or sufficient filling to stuff the cushion.

If preferred, make a cardboard template of the shape measuring 5 cm (2 in) square: giving a finished square of 3.8 cm (1½ in) plus a 5 mm (¼ in) seam allowance all round.

Cut the fabric into the following numbers of 5 cm (2 in) squares.

PN × 16
PP × 18
LF × 34
SF × 32

Reserve a square of backing material approximately 40 cm (16 in) square. If the cushion is to be piped, also cut sufficient bias strips 3.8 cm (1½ in) wide from the chosen fabric to give a final strip of 152.5 cm (61 in) in length. Bias strips are cut diagonally across the grain of the fabric so that they will stretch when used.

Sew square pieces together along the seam allowance in the order shown on the layout, making strips each of ten squares and then joining them, carefully matching the seams.

If the cushion is to be piped, join the bias pieces by hand or machine into a strip 3.8 × 152.5 cm (1½ × 61 in), then make the piping by doubling the bias fabric right side out, along its length, and folding the cord into the centre of the strip. Machine along the length of the folded bias strip, near to the enclosed cord, so stitching the cord inside the fabric and creating piping.

Trim the backing fabric to match the completed patchwork front (approximately 38 cm/ 15 in square) and, with right sides facing seam them together around three sides by hand or machine along the seam allowance. Turn right side out, insert the cushion pad or filling and, as invisibly as possible, close the fourth seam with oversewing.

If using piping, it should be inserted within all these seams, allowing only the covered cord to show on the right side of the seam, and stitching the raw edges of the bias strip into the seam between the back and front pieces.

Finished size: 38 cm (15 in) square.

TRADITIONAL PATCHWORK CUSHION***

Still based on straight lines, this is a more complex pattern of interlocking rectangles known as Log Cabin. The pattern suggests the arrangement of logs in early houses and is traditionally worked in a mixture of dark and pale colours, the dark making one half of each motif, the pale the other. The square motifs are then assembled so that the dark corners are together.

This cushion is also piped, but it can be made without piping. One side of the cushion cover could be closed with a zip or other fastening for easy removal.

Fabric code
M = plain maroon
DF = dark floral (large flowers on navy ground)
GF = green floral (very small green floral on cream ground)
PF = pink floral (very small pink floral on cream ground)

The wooden pulls on blind strings are acorn shaped because the oak tree is sacred to thunder, and will never be struck by lightning. These wooden acorns therefore protect the windows of the house from this unlikely mishap.

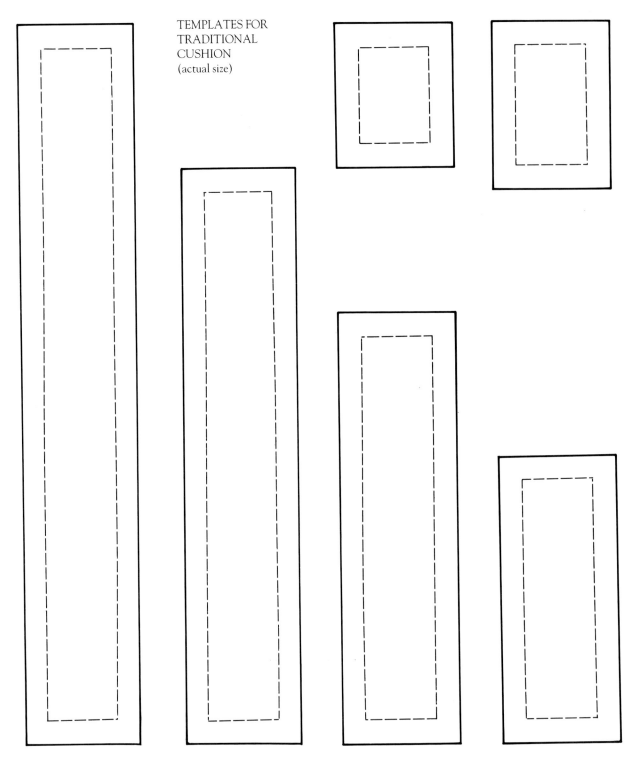

TEMPLATES FOR
TRADITIONAL
CUSHION
(actual size)

For each of the above, 1 piece approximately 25 × 41 cm (10 × 16 in) is required.
N = navy pattern 1 piece 7.5 × 7.5 cm (3 × 3 in)
For the backing: 1 piece 41 cm (16 in) square.

For piping (if required): sufficient to cut a bias strip 3.8 × 145 cm (1¹/₂ × 57 in)

If required, approximately 145 cm (57 in) of piping cord

Cushion pad approximately 36 cm (14 in) square, or sufficient filling to stuff the cushion.

If you prefer, make cardboard templates in the following sizes (the sizes include a 6 mm (¹/₄ in) seam allowance all round):
19.3 cm (7¹/₂ in) × 3.2 cm (1¹/₄ in)
15.3 cm (6 in) × 3.2 cm (1¹/₄ in)
11.5 cm (4¹/₂ in) × 3.2 cm (1¹/₄ in)
7.6 cm (3 in) × 3.2 cm (1¹/₄ in)
4.5 cm (1³/₄ in) × 3.2 cm (1¹/₄ in)
3.2 cm (1¹/₄ in) × 3.2 cm (1¹/₄ in) for the central square.

Cut the fabric into the following number of strips, using the templates if preferred:
M 8 × 15.3 cm (6 in) strips
   8 × 7.6 cm (3 in) strips
DF 4 × 19.1 cm (7¹/₂ in) strips
   8 × 11.5 cm (4¹/₂ in) strips

4 × 4.5 cm (1³/₄ in) strips
GF 4 × 19.1 cm (7¹/₂ in) strips
   8 × 11.5 cm (4¹/₂ in) strips
   4 × 4.5 cm (1³/₄ in) strips
PF 8 × 15.3 cm (6 in) strips
   8 × 7.6 cm (3 in) strips
N 4 × 3.2 cm (1¹/₄ in) squares

Also, reserve a square of backing fabric with approximately 35 cm (15 in) sides. If the cushion is to be piped, also cut sufficient bias strips 1¹/₂ in (3.5 cm) wide, from the chosen fabric, to give a final strip of 145 cm (57 in) long.

Using the layout as a guide, assemble four separate log cabin motifs in the colours stated, sewing along the seam allowances, matching corners carefully and keeping seams square.

When all four motifs are completed, assemble into the following pattern, known as Court House Steps, rotating each motif through 90° so that all dark corners are in the centre. On the layout, the strips marked * are the DF strips.

Work piping and backing as given in the instructions for the Modern Patchwork Cushion (page 88), noting that this cushion is only 36 cm (14 in) square.

Finished size: approximately 35 cm (14 in) square.

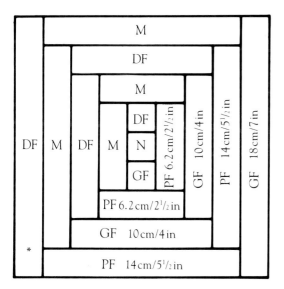

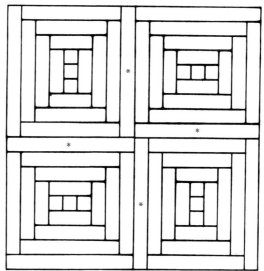

CUSHION LAYOUT
Layout for single log cabin motif (far left) (finished size), and Court House steps pattern (left)
DF = 3.2 cm/1¹/₄ in
     × 1.9 cm/³/₄ in
N = 1.9 cm/³/₄ in
     × 1.9 cm/³/₄ in
GF = 3.2 cm/1¹/₄ in
     × 1.9 cm/³/₄ in

BABY'S COT QUILT * * *

This pattern, called Mosaic, uses hexagon shapes grouped in flower patterns, with a patch of one colour at the centre surrounded by six of another colour. These groups join together in a repeating mosaic. Floral fabrics were used here, but you can use any material you wish.

Fabric code
W = white with very small pattern
112 × 130 cm (44 × 51 in)
DB = darker blue with feather pattern
88 × 40 cm (35 × 16 in)
PB = paler blue flowers on white 88 × 40 cm
(35 × 16 in)
DP = darker pink floral on cream 88 × 40 cm
(35 × 16 in)
PP = paler pink floral on white 88 × 40 cm
(35 × 16 in)
One piece of quilt wadding at least 115 × 86 cm
(46 × 34$\frac{1}{2}$ in)

You can use the diagrams to make a cardboard template or trace a paper pattern of each of the three shapes.
Using the templates or patterns, cut the fabrics into the following pieces:
W = 30 hexagons, 12 triangles.
DB = 31 hexagons, 2 × $\frac{1}{2}$ hexagons.
PB = 25 hexagons, 7 triangles, 3 × $\frac{1}{2}$ hexagons.
DP = 26 hexagons, 4 × $\frac{1}{2}$ hexagons.
PP = 25 hexagons, 7 triangles, 3 × $\frac{1}{2}$ hexagons.
Also reserve a piece of **W** at least 118 × 89 cm
(47 × 35$\frac{1}{2}$ in) for the backing.

Sewing along the seam allowances, stitch the pieces together in the order shown on the layout. When using a sewing-machine, it is usual to stitch the hexagons together in rows and then join the rows with a zig-zag seam. By hand, it is probably easier to stitch a 'daisy' of hexagons, and then join these daisies together. Finally, sew in the irregularly shaped pieces around the edge.
   Trim the backing sheet of W fabric to match the patchwork front. Placing right sides together,

A delightful cot quilt can be made quite easily using a pattern of hexagon patches.

join the backing and the patchwork together by hand or machine around three sides, using the seam allowances as before. Turn the right side out. Trim the wadding if necessary so that it is about 2.5 cm (1 in) smaller all around than the finished quilt cover, then carefully place the wadding inside the cover, ensuring that it is completely flat. Oversewing as invisibly as possible, close the fourth and final edge of the cover, so containing the wadding inside.

Note: Take great care to ensure that the pieces of backing fabric and wadding are cut exactly on the square in order to keep the quilt in shape.

Finished size: approximately 115 × 86 cm (46 in × 34¹/₂ in).

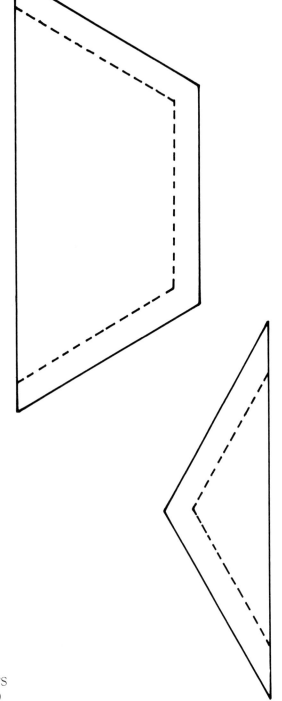

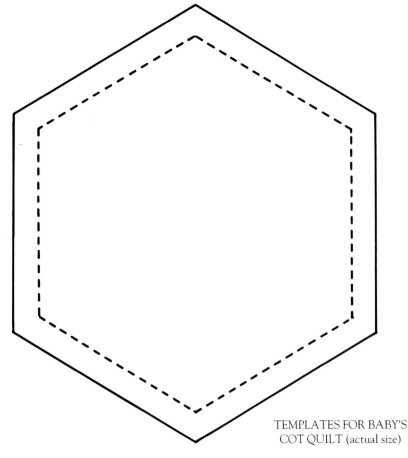

TEMPLATES FOR BABY'S
COT QUILT (actual size)

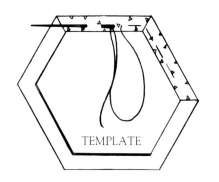

TEMPLATE

JOINING HEXAGONS

◀ Turn seam allowances very accurately onto wrong side and tack down. This can be done over a very accurate hexagon of stiff paper, which is later removed and can be re-used.

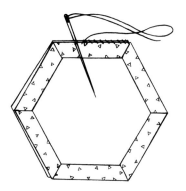

◀ With right sides facing, and with the corners exactly matched up, neatly oversew the hexagons together.

LAYOUT FOR BABY'S COT QUILT (below)

ASSEMBLING THE HEXAGONS

◀ Make separate daisies from seven hexagons

▶ Join the daisies together

◀ Fill in the edges with part daisies and hexagons

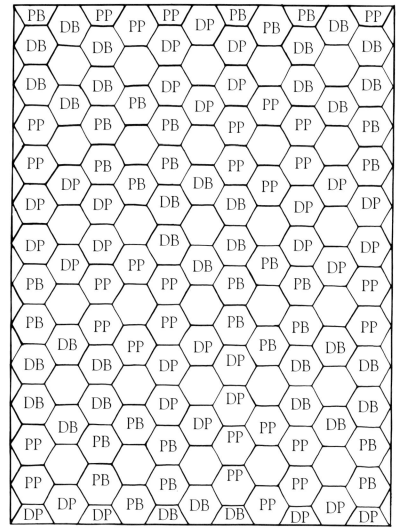

RIBBON CARDS *

This is a very easy way of making most attractive cards. They only just count as patchwork, but they belong here because they consist of an arrangement of fabric making a pattern. Two have been made: one an extremely simple arrangement of diagonal lines of Christmas ribbon; the other a more elaborate, but still easy, woven pattern of ribbons with motifs. The technique used is a simple and effective way of securing ribbon patterns, and can be used in all sorts of ways. Card blanks with a window cut out of the front are needed, and can be found at a craft shop. You can make your own but they are quite difficult to do. If using a home-made card, make sure that it is cut and folded very accurately so that it stands well. Be sure also that, unless otherwise preferred, it opens the 'correct' way, that is, like a book; and that it will fit an available envelope.

▲

Card blank with a window hole in the front
Lengths of matching or toning fabric ribbon
which can be ironed (do not use plastic or florists
ribbon, which is not heat resistant)
Piece of iron-on interfacing
Piece of thin card (optional)
Paper or fabric glue

▼

Plan your pattern. The ribbons are going to be arranged or woven into a pattern that will fill the window in the card, with some overlap on the wrong side for attaching them to the card. The Christmas ribbon is simply arranged in parallel stripes; the motif ribbon is woven with two plain colours alternating with two motif ribbons.

Cut the ribbon into lengths suitable for your pattern, ensuring that the pieces are long enough to make an area big enough to fill the window in the card, with about an extra 2.5 cm (1 in) at each end for pinning and working.

A padded surface is needed into which you can stick pins: either a padded ironing board or a folded towel or blanket covered with a sheet or cloth. The surface needs to be smooth as any

texture will become pressed into the ribbons when they are ironed. On the padded surface, arrange the ribbons of your choice face down. You will find that the ribbon is sufficiently fine for you to see the motifs through it and so to arrange the pattern as you wish. In this case, a central spot features in each square of the woven pattern, and the heart motifs on the Christmas ribbon face each other. Pin through the ribbons into the padded surface to hold them in position. If ribbons are to be woven, arrange all the ribbons which go in one direction, the 'warp', first, and pin them down at one end only. Then weave the ribbons which are to go in the opposite direction, the 'weft', through the warp ribbons at the free ends of the warp ribbon and slide them up to the other end. Secure the weft ribbons at both ends by pinning into the padded surface. Finally pin down the free ends of the warp ribbons. Cut a piece of iron-on interfacing to the shape of the window in the card, but at least 1 cm (½ in) larger all round. Following the instructions which come with the interfacing, iron it onto the wrong side of the ribbons as they are pinned down.

When cool, the pins can be removed from the ribbon and it can be handled as though it were one piece of fabric. Cut to size a little larger than the window in the card, position it in the window on the inside of the card and glue it down. If preferred, a piece of plain card can then be glued over the back of the ribbons to cover them.

Apple pie wi' out t'cheese
Is like a kiss wi' out a squeeze.

Embroidery

Embroidery is an easy skill that will enable you to decorate all sorts of fabric items such as bookmarks and tablecloths with pretty designs, or to make a tapestry picture of a favourite scene.

O ne of the most gentle of crafts, embroidery is enjoying a big comeback at the moment and all kinds of needlework are popular again. Historically it was used to decorate all kinds of fabrics, rich dresses, wallhangings, church vestments and household linen, and all of these provide ideas for modern designs. The very simple bookmark could be worked by a beginner, but the tapestries need a little more experience. The tablecloth is based on the traditional format of a pattern at each corner with a linking motif, brought up to date with designs which take their inspiration from Yorkshire. Embroidery is noted for its therapeutic properties and can be most relaxing and soothing. The results, as they decorate tables and other items around the home, are certainly very rewarding.

TAPESTRY BAG***

The principle of making a bag from tapestry has a long tradition. People have always liked small tapestry purses, and handbags and even carpet bags have long been made in this way. Pure wool tapestry is soft but strong, and tough enough to survive the hard wear which it is likely to encounter. This very simple bag is an envelope of tapestry lined with fabric and the same principle could be used to make a purse or bag of any size. The pattern is an all-over floral repeat, again a carpet-bag idea; but if preferred just one motif could be worked, perhaps on the front of the bag, the rest being worked in the plain ground colour.

▲

Patons Pure Wool Superwash Double Knitting × 50 gm balls:
1 ball ground colour (cream);

plus, part balls of purple-blue, dark pink, pale pink, pale yellow, dark yellow, paler green and darker green.
10 holes per 2.5 cm (1 in) tapestry canvas at least 35 × 65 cm (14 × 26 in).
Piece of lining material 32.5 × 63 cm (13 × 25 in).
Press fasteners (optional).

▼

It is very important that this piece of tapestry work comes out 'square' in order that the bag will be a true rectangle. The work will keep its shape best if worked in a tapestry frame.

Mark out a central area 30 × 60 cm (12 × 24 in) and therefore 120 holes × 240 holes on the canvas. Work within this area, starting along one shorter side, which becomes the bottom edge of the work. Working from the colour chart on p. 100, in half cross stitch:

*Starting at A as the first stitch at the bottom righthand corner, work the 40-stitch repeat three times in all across the work. Using this as an establishing row of stitches, work all three repeat motifs from the chart until they are complete, so working 40 lines of stitches in all.

On the next line of stitches begin at the righthand side with the stitch marked B, and work across the first line of stitches of the lefthand half of the motif, then two more whole motifs, finishing with the righthand half of the motif. Using this as an establishing row of stitches, work these two halves and two whole repeat motifs until they too are complete. *

Repeat from * to * twice more, so working a total of 6 rows of motifs, which alternate in the way in which they are arranged from side to side. The completed area of canvas worked will be approximately 30 × 60 cm (12 × 24 in).

If the work is not a true rectangle, pin it into shape on a firm padded surface (a carpeted floor is ideal) and press under a damp cloth with a very warm iron. Leave to dry before unpinning.

Trim the canvas to within about 1 cm (½ in) of the stitches all around. Turn the canvas edges neatly to the wrong side and catch down invisibly to the back of the work with a hemming stitch.

The tapestry bookmark and evening bag demonstrate the versatility of this popular and easy needlecraft.

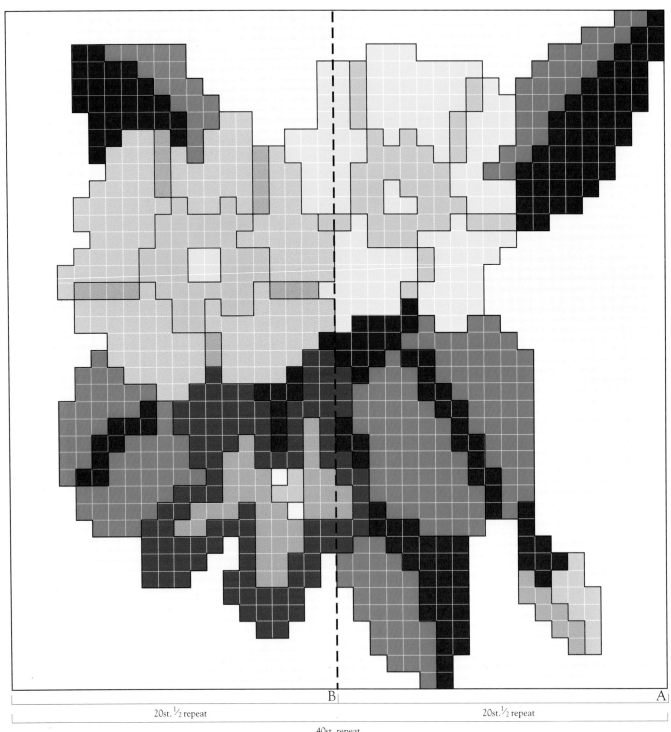

B
A

20st. ½ repeat

20st. ½ repeat

40st. repeat

Turn the hem allowance of the lining fabric onto the wrong side so that this finished rectangle is just smaller than the finished canvas work.

Tack these hems in place. Neatly and as invisibly as possible attach the lining in position, wrong sides together, stitching it on with a hemming stitch all around its hemmed edge just inside the hemmed edge of the tapestry work. Remove the tacking stitches from the lining fabric.

Fold the work into thirds with the lining inside, so creating an envelope bag 20 × 30 cm (8 × 12 in) with a closing flap which completely covers the front.

Oversew the sides using the cream wool and working very neatly because the stitches will be visible, and leave the front flap free.

The front flap can be closed with press fasteners.

Finished size: approximately 20 × 30 cm (8 × 12 in)

ASSEMBLING THE TAPESTRY BAG

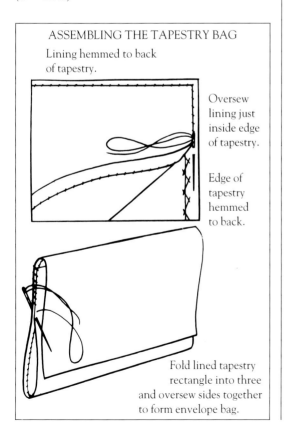

Lining hemmed to back of tapestry.

Oversew lining just inside edge of tapestry.

Edge of tapestry hemmed to back.

Fold lined tapestry rectangle into three and oversew sides together to form envelope bag.

SHEEP BOOKMARK *

This bookmark is an easy cross stitch project. If you are new to the craft, use this piece to practise crossing all the stitches the same way and keeping the fabric square and even. To finish the bookmark and prevent the edges fraying, you can either back it with iron-on interfacing, or fold the sides to the back to make it double thickness.

To fasten on a piece of yarn, leave an end of about 2.5 cm (1 in) free at the back of the fabric, and catch it down at the back with every stitch you make as you go along. Work the crosses of thread over the canvas in the pattern described below, ensuring that they all appear the same and cross in the same direction. Fasten off after a complete stitch by passing the thread under stitches on the back of the work before cutting off the end.

▲

1 piece of cross stitch fabric, 11 holes per 2.5 cm (1 in), at least 7.5 × 25 cm (3 × 10 in) if using iron-on interfacing; at least 15 × 25 cm (6 × 10 in) if not.
1 (optional) piece of iron-on interfacing at least 7.5 × 23 cm (3 × 9 in).
Coats Nordin Cross Stitch threads: 1 skein each of white, dark brown and green.

▼

Beginning at the bottom of the material, mark the centre thread 5 cm (2 in) up from the bottom edge.

Using this mark to place the pattern, and working from the chart, work one complete white sheep motif starting with line 1, with the stitch marked * as the 10th stitch to the right of the centre.

When the 14 lines of the white sheep are completed, leave 4 lines of thread without any embroidery, then work a black sheep motif, locating it so that the centre stitch and the stitch marked * are directly above those similarly marked on the white sheep.

When the 14 rows of the black sheep are completed, work two further white sheep, each beginning 4 thread lines above the previous motif and each located centrally as before.

STITCH CHART FOR THE SHEEP BOOK-MARK

14 lines of white sheep motif

4-line space

14 lines of black sheep motif

KEY
○ = white
✕ = dark brown
\ = green
L = white lines

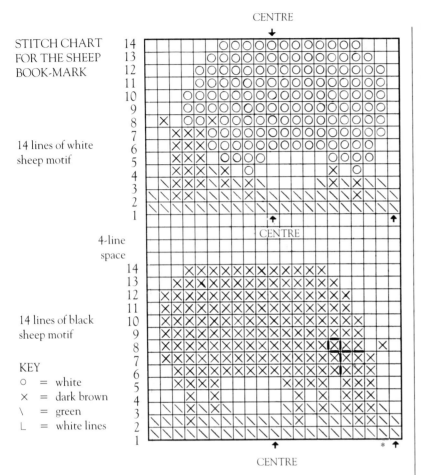

CENTRE

CENTRE

CENTRE

To complete the bookmark, outline the white area of each white sheep with a single embroidered dark brown line worked in backstitch. Embroider a single white line where shown on the chart on the black sheep, also in backstitch.

If using iron-on backing, attach it carefully to the back following the manufacturers instructions. Cut to shape using fabric threads as a guide. The finished shape is approx 6 cm (2½ in) wide by 23 cm (9 in) long, tapered at the lower end.

If preferred, double the fabric at the sides to the back and join in a centre back-seam in backstitch. Turn hems in at the top edge and on the shaped bottom edge and oversew as invisibly as possible.

Decorate the finished bookmark by stitching on a tassel of embroidery thread at the bottom point.

Finished size: approximately 6 × 23 cm (2½ × 9 in)

This tapestry picture of Emmerdale is worked in half cross stitch.

TAPESTRY PICTURE OF EMMERDALE***

Emmerdale is epitomized for most people by the silhouetted outline of a peaceful Dales farmhouse surrounded with trees and grazing cattle.

The picture is worked in half cross stitch, which is simply a series of straight diagonal stitches each of which is worked over one cross of the threads of the canvas. Fasten on either by knotting the tapestry wool, or by leaving a short end which is then caught down behind the work as the stitches are made. Fasten off by threading the last end under some of the stitches on the back.

It is most important that all of the diagonal stitches on the front of the work travel in the same direction to achieve a smooth and attractive finish. That is, they should all be from the same bottom corner to the same top corner.

▲

Coats Anchor Tapisserie Wool in 10 metre (33 ft) skeins in the following colours:
1 skein each of: 0402, 0403, 0987, 0360, 3041, 0984, 0391, 0399, 0397, 0263, 0266, 3087, 0268, 3236, 0733, 0160, 0386;
2 skeins each of: 3234, 0568, 0736.
1 piece of tapestry canvas at least 26 × 40 cm (10½ × 16 in), with 10 holes or threads per 2.5 cm (1 in).
Tapestry frame if available.
Optional picture frame and glass to fit, and a mount with a picture oval approximately 25 cm (10 in) in width.

▼

Following the colour chart on pp. 104–5, work the tapestry in half cross stitch as shown in the centre of the canvas. On the chart, one square represents one stitch. Take care that the canvas distorts as little as possible; the best way is to work with the canvas in a tapestry frame.

When completed, mount and frame the tapestry, or have it professionally framed.

Finished size: approximately 26 cm (10½ in) across.

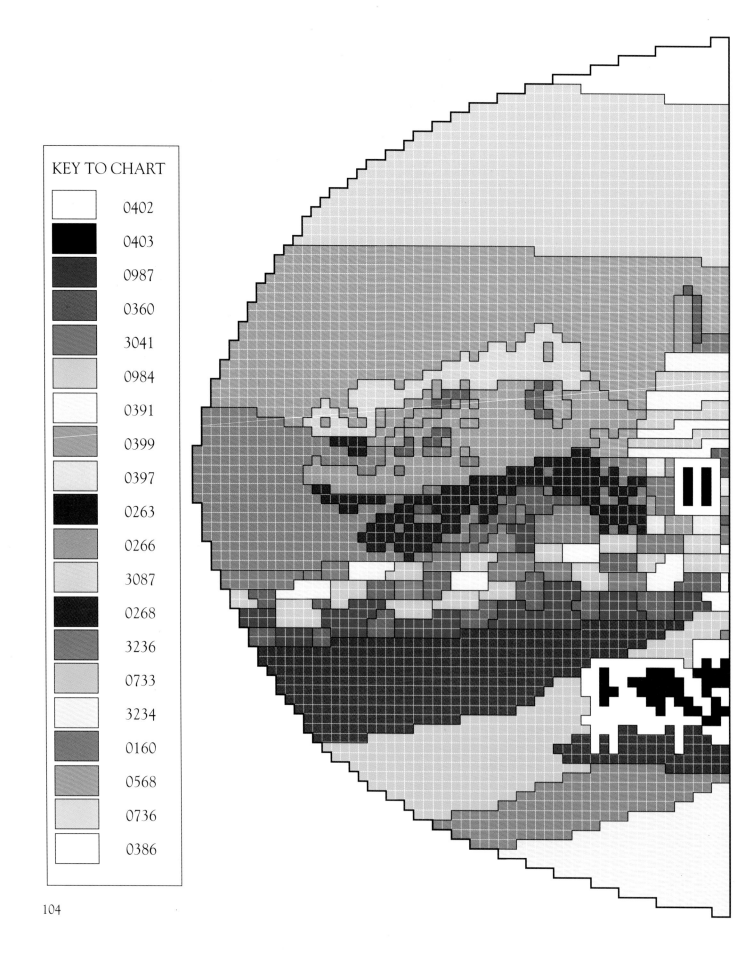

KEY TO CHART

	0402
	0403
	0987
	0360
	3041
	0984
	0391
	0399
	0397
	0263
	0266
	3087
	0268
	3236
	0733
	3234
	0160
	0568
	0736
	0386

EMMERDALE CORNFIELD TABLECLOTH***

As a child I always thought that, if ever I owned a farm, I would sow poppy and cornflower seed with the wheat. The scarlet and blue look beautiful in the gold wheat just before harvest, and I have loved the colours ever since. Today, if you look at the headlands of Yorkshire cornfields, you can still see these flowers under and around the dry stone walls, so here they are, embroidered on this cloth.

The silks used are the six-stranded kind, halved to give three-stranded working threads. The stitches are simple and involve mainly the use of straight stitches of various lengths as shown in the diagrams. French knots are worked by bringing the needle through from the back, winding the silk a few times around the needle, and returning it back through the fabric, pulling the knot firm. Take care not to pull any stitches too tight, as this can distort the work and spoil the effect.

To fasten on the thread to start embroidering, a simple knot can be made in the end of the thread which will not pull through the fabric. Two or three oversewn stitches on the back of other stitches will serve to fasten off.

▲

A piece of suitable fabric, e.g. linen, at least 88 × 88 cm (35 × 35 in), or a plain, made-up cloth.
Coats stranded embroidery cotton (silks), the following number of 8 m (27 ft) skeins:
poppy red 3 skeins
black 4 skeins
1 skein of each of the following:
darker cornflower blue, paler blue, yellow, brighter green, darker green, gold, white, darkest grey, mid grey, lightest grey
Embroidery tracing pencil
Tracing paper

▼

Cut an accurate 88 × 88 cm (35 × 35 in) square from the linen, if not using a ready-made cloth.

Using the layout on page 109 as a guide, mark the positions of all the motifs. The floral motif is worked four times and is set within the area of a 15 cm (6 in) diameter circle; each motif is positioned at a corner, 10 cm (4 in) in from the sides of the cloth. Two wall motifs are spaced on each side between these floral motifs with a 5 cm (2 in) central gap as shown.

Using the embroidery tracing pencil and the tracing paper, and following the instructions which accompany the pencil, trace the floral motif on page 108 four times and the wall motif on page 109 eight times onto the cloth in the positions marked. Make as few marks on the cloth as you need in order to work the embroidery.

Work the floral and wall motifs in embroidery using the guides on pages 108 and 109.

If you are using a piece of linen, turn a 1 cm (¹/₂ in) hem to the wrong side and sew by hand or machine, to produce an 85 cm (34 in) square.

Finished size: 85 × 85 cm (34 × 34 in)

Note: When working individual motifs, it is easier to keep the work flat and in shape if an embroidery frame is used. The finished cloth should be carefully pressed on the wrong side on a well padded surface.

COLOUR GUIDE

Tablecloths can be embroidered to create unique and treasured items. This cornfield motif is a simple one to begin with.

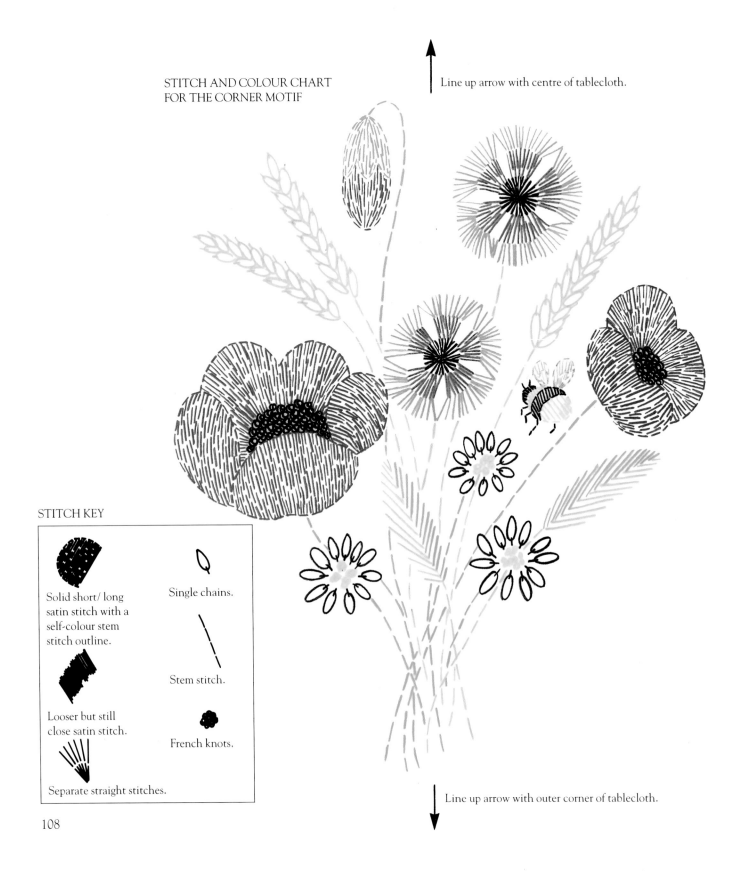

STITCH AND COLOUR CHART
FOR THE CORNER MOTIF

Line up arrow with centre of tablecloth.

STITCH KEY

Solid short/ long satin stitch with a self-colour stem stitch outline.

Single chains.

Stem stitch.

Looser but still close satin stitch.

French knots.

Separate straight stitches.

Line up arrow with outer corner of tablecloth.

STITCH GUIDE FOR THE WALL MOTIF

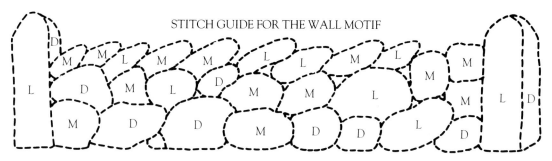

Gateposts are lined as shown in colour indicated.

--- = lines of stem stitch in black

Each stone is then partly filled with long and short satin stitch in the colour indicated:
D ´ = darkest grey
M = mid grey
L = lightest grey

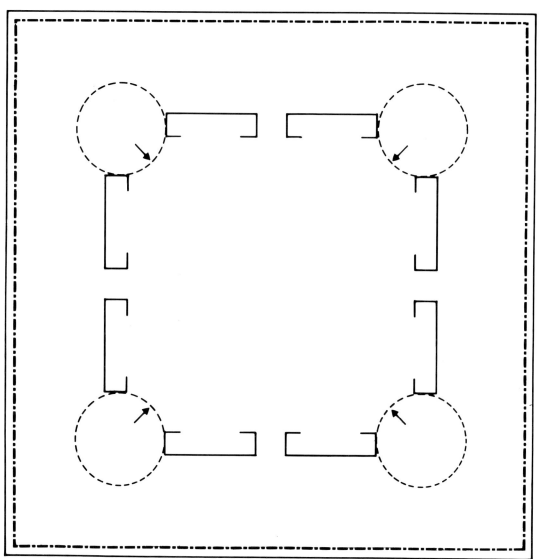

TABLECLOTH POSITION GUIDE

—·—·—·—
hemline

position of floral motif ×4, with top as arrowed

position of wall motif ×8

Crochet

Mastery of this simple craft will enable you to create delicate, lacey material that can be used to enhance clothes and all kinds of household linen.

Crochet is a form of quickly-made lace, and looks at its best when it is used over another surface such as linen, wood or even window glass. It is firm and tough, making it ideal for decorating things which will have to be washed regularly. The small items here are also very economical to make.

Traditionally crochet was to be found anywhere in a country home where a little decoration was needed. These patterns carry on the tradition. Crisp cottons have been used because they suit the craft, and wash and wear for ever. In collections of historic costumes far more crochet survives than knitting, mainly because it is so tough, so do not be afraid to use it for things like bed linen which need a lot of washing.

Note: Crochet worked in cotton looks best if pressed under a damp cloth after being pinned or blocked into shape and size.

ABBREVIATIONS

UK	AMERICAN
ch = chain	
(dc) double crochet	– (sc) single crochet
(tr) treble crochet	– (dc) double crochet
(dtr) double treble	– (tr) treble crochet
(htr) half treble	– (hdc) half double crochet
(inc) increase	
	(dec) decrease
	(sl st) slip stitch
	(sp) space
	(st) stitch
	(tog) together
	(yo) yarn over

Crochet can be used to create edging and trims in a variety of patterns for all sorts of items, such as the wave pattern on the child's nightdress and the diamond pattern on a pillowcase.

EDGING FOR A CHILD'S NIGHTDRESS**

This very beautiful wave edging would lull any little girl happily to sleep, and with a small amount of yarn and lots of love you can transform an ordinary nightdress into an heirloom.

▲

Coats Mercer Crochet Cotton, No. 20 in ecru: approximately 25 gm
1 mm hook.

▼

Tension: 25 tr sts and 9 rows of trs = 5 cm (2 in) worked on a 1 mm hook.

WAVE PATTERN
Make 8 ch and join into a ring with a sl st into first ch.
Row 1 – 3 ch, 8 tr into ring, turn.
Row 2 – 4 ch, 1 tr into 2nd tr, * 1 ch, 1 tr into next tr, rep from * 5 times more, 1 ch, 1 tr into 3rd of 3 ch.
Row 3 – 5 ch, 1 tr into 2nd tr, * 2 ch, 1 tr into next tr, rep from * 5 times more, 2 ch, 1 tr into 3rd of 4 ch.
Row 4 – 6 ch, 1 tr into 2nd tr, * 3 ch, 1 tr into next tr, rep from * 5 times more, 3ch, 1 tr into 3rd of 5 ch.
Row 5 – * (1 dc, 3 tr, 1 dc) into 3 ch sp, rep from * 7 times more, 8 ch, turn and sl st into 2nd of first 3 tr, turn.
Row 6 – 3 ch, 8 tr into 8 ch loop, turn.
Row 7 – 4 ch, 1 tr into 2nd tr, * 1 ch, 1 tr into next tr, rep from * 5 times more, 1 ch, 1 tr into 3rd of 3 ch, sl st into 2nd tr of next group of 3, turn.
Row 8 – 5 ch, miss first tr, 1 tr into next tr, work as row 3 from * to end.
Row 9 – Work as row 4, then sl st into 2nd tr of next group of 3, turn.
Rep rows 5 to 9 for the length required, then work row 5, omitting the 8 ch at end of row.

HEADING
Working along straight edge of work, 5 ch, 1 tr into next row-end, * 2 ch, 1 tr into next row-end, rep from * to end, turn.

Next row – 3 ch, * 2 tr into 2 ch sp, 1 tr into next tr, rep from * to end, working last tr into 3rd of 5 ch.
Fasten off.

TO MAKE UP

Work pattern to a sufficient length ending with a complete motif:
For the cuffs, work to a length which will cover the cuff, here 8 repeats for 25 cm (10 in).
For the collar work 2 pieces each to a length to cover half the collar, here 6 repeats were required for each 18 cm (7 in) half.

Work to a greater rather than lesser length when working to the nearest repeat. The edging looks prettier gathered rather than stretched.

To attach the collar, reverse half of the collar, to give two symmetrical pieces. Each piece will be attached around one half of the neck opening. Stitch on all the pieces as invisibly as possible.

Finished size: The edging is approximately 3.7 cm (1³/₄ in) deep and used here were: collar, 36 cm (14 in) long; 2 cuffs, each 25 cm (10 in) long.

STITCH CHART
FOR PILLOW-CASE
EDGING

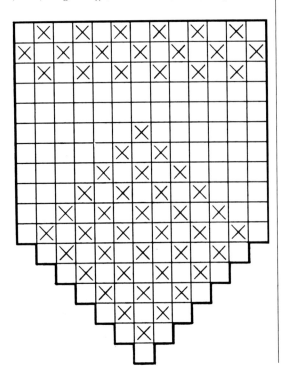

EDGING FOR PILLOW-CASE**

This is another traditional use for crochet, making an ordinary pillow-case into something really luxurious for a guest room. Rather than adding a narrow decorative edge, this rich, flat crochet lace covers a wide band of the fabric itself.

▲

Coats Mercer Crochet Cotton No. 5 in white:
approximately 60 gm
1.75 mm hook

▼

Tension: 10 sts and 14 rows of pattern = 10 cm (4 in) worked on a 1.75 mm hook.

POINTS PATTERN

Make 38 ch.
Row 1 – 1 tr into 8th ch from hook, * 2 ch, miss 2 ch, 1 tr into next ch, repeat from * to last 6 ch, 1 tr into each of next 3 ch, 2 ch, miss 2 ch, 1 tr into next ch, 3 ch, turn.

Row 2 – 2 tr into 2 ch sp, 1 tr into next tr, 2 ch, miss 2 tr, 1 tr into next tr, 2 tr into 2 ch sp, 1 tr into next tr (2 ch, 1 tr into next tr) 7 times, 2 tr into 2 ch sp, 1 tr into 5th of 7 ch, 6 ch, 1 dtr into base of last tr, sl st, into each of last 3 ch (1 sp increased), 7 ch, turn.

Row 3 – 1 tr into last sl st (1 sp increased), 2 tr into 2 ch sp, 1 tr into next tr, 2 ch, miss 2 tr, 1 tr into next tr, 2 tr into 2 ch sp, 1 tr into next tr (2 ch, 1 tr into next tr) 6 times, 2 ch, miss 2 tr, 1 tr into next tr, 2 tr into 2 ch sp, 1 tr into next tr, 2 ch, miss 2 tr, 1 tr into 3rd of first 3 ch, 3 ch, turn.

Work in pattern to a sufficient length, ending with a complete repeat. Here, 6 repeats are worked. It is better to work to a greater rather than lesser length, as it looks better slightly gathered than stretched. Fasten off.

To attach the edging, stitch on evenly and as invisibly as possible, by the long straight edge and the two ends only.

Finished size: The edging is approximately 15 cm (6 in) deep, and each piece here is approximately 45 cm (18 in) long to fit a standard pillow-case.

MATS AND COASTERS**

This is a familiar use of crochet. Hexagon shapes are ideal because some can be used individually, and others can be sewn together to create a centre mat of whatever size you want. Seven hexagons were used here to make a traditional table centre, but you can use as many as needed for the size you require.

▲

Coats Mercer Crochet Cotton No. 10 in cream: approximately 40 gm. 1.25 mm hook.

▼

Tension: 14 tr sts and 8 rows of trs = 5 cm (2 in) worked on a 1.25 mm hook.

SPECIAL ABBREVIATIONS

Tr 4 tog and tr 5 tog = * yo, insert hook as indicated, yo, draw loop through, yo, draw through 2 loops * rep. from * to * for each 'leg' of the cluster. = 5(6) loops on hook end, yo, draw through all loops on hook.

Make a total of 11 hexagons; 4 are used as coasters. To make 1 hexagon:
Base ring: 6 ch, join with sl st.
1st round – 1 ch, 12 dc into ring, sl st to first dc (12 sts).
2nd round – 1 ch, 1 dc into same place as 1 ch, (7 ch, miss 1 dc, 1 dc into next dc) 5 times, 3 ch, miss 1 dc, 1 dtr into top of first dc.
3rd round – 3 ch (count as 1 tr), 4 tr into arch

The polished wooden surface perfectly sets off the crocheted pattern of the coasters and table mat.

formed by dtr, (3 ch, 5 tr into next 7 ch arch) 5 times, 3 ch, sl st to top of 3 ch.

4th round – 3 ch (counts as 1 tr), 1 tr into each of next 4 tr, * 3 ch, 1 dc into next 3 ch arch, 3 ch * *, 1 tr into each of next 5 trs; rep from * 4 more times and from * to * * again, sl st to top of 3 ch.

5th round – 3 ch, tr 4 tog over next 4 trs (counts as tr 5 tog), * (5 ch, 1 dc into next 3 ch arch) twice, 5 ch * *, tr 5 tog over next 5 trs; rep from * 4 more times and from * to * again, sl st to first cluster.

6th round – sl st into each of next 3 ch, 1 ch, 1 dc into same place, * 5 ch, 1 dc into next 5 ch arch; rep from * all round omitting last dc and ending sl st to first dc.

7th round – sl st into each of next 3 ch, 1 ch, 1 dc into same place, * 5 ch, 1 dc into next 5 ch arch, 3 ch, (5 tr, 3 ch, 5 tr) into next arch, 3 ch, 1 dc into next arch; rep from * 5 more times omitting last dc and ending sl st to first dc. Fasten off.

To make up the mat, use one hexagon as the central piece and join the remaining 6 each by one edge to one edge of the central hexagon. Join these outer hexagons to each other.

Finished size: One hexagon (i.e., one coaster) = 7.5 cm (3 in) across; 7 hexagons assemble into a mat approximately 23 cm (9 in) across.

WINDOW INSET WITH DIAMOND PATTERN***

Windows in a Yorkshire Dales farmhouse tend to be made up of small panes. Window sills provide space, leaving little room for net curtains. Instead, these crochet insets can be used.

▲

Coats Mercer Crochet Cotton No. 5 in white: approximately 30 gm for each curtain.
3 mm hook.

▼

Tension: 11 sts and 14 rows of tr mesh pattern = 10 cm (4 in) worked on a 3 mm hook.

Make 66 ch.

First row – 1 tr into 4th ch from hook, 1 tr into each ch to the end. 21 blocks each of 3 trs.

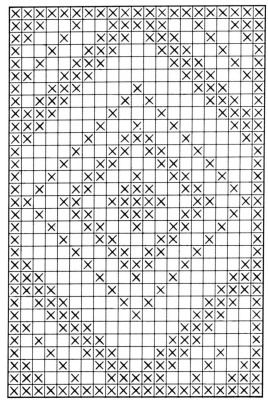

STITCH CHART FOR WINDOW INSET

2nd row
1st row

2nd row – 3 ch as first tr, 1 tr into next 8 trs, 2 ch miss 2 trs, 1 tr into next tr, 2 ch miss 2 trs, 1 tr into next tr, 1 tr into next 4 trs, 2 ch miss 2 trs, 1 tr into next tr, 2 ch miss 2 trs, 1 tr into next 16 trs, 2 ch miss 2 trs, 1 tr into next tr, 2 ch miss 2 trs, 1 tr into next 4 trs, 2 ch miss 2 trs, 1 tr into next tr, 2 ch miss 2 trs, 1 tr into next 9 trs.

Using these as establishing rows work from the stitch guide. Each X within the guide represents a block of 3 trs, and each unmarked square an open trellis. Continue until chart is completed and all 31 rows have been worked.
Fasten off.

To finish the window inset, stretch and fasten to a window-pane frame using small, large-headed tacks or drawing pins. Short pins with coloured heads in white or the colour of the curtain would be inconspicuous.

Finished size: approximately 21 × 24 cm (8¹/₂ × 9¹/₂ in) unstretched.

Crochet window insets make a delightful and unusual alternative to net curtains.

115

Knitting

Knitting is a skill that is quickly acquired, and you can use it to make warm rugs and socks, traditional cushion covers, or a special bedspread.

The people of the Dales have always knitted, either out of economic necessity or to create something beautiful to wear or for the home. Few people now need to knit, but many like to keep alive the tradition of creating decorative household items using the most simple of all household crafts, knitting. These patterns are traditional without being old fashioned, and you can work a simple rug, traditional socks, more complicated aran patterns, or tackle that labour of love, an heirloom bedspread. Like embroidery, knitting is very therapeutic and many people find it the most relaxing and compulsive craft of all.

ABBREVIATIONS
(These will be the same for all the knitted designs)
K = knit; P = purl; st(s) = stitch(es); st st = stocking stitch; inc = increase; dec = decrease; beg = beginning; cont = continue; foll = follows(ing); tbl = through back of loop; sl = slip; psso = pass slip stitch over; yon = yarn over needle; yrn = yarn round needle; rem = remaining; tog = together; patt = pattern.

AMERICAN AND BRITISH KNITTING TERMS

UK		US
stocking stitch	=	stockinette stitch
tension	=	gauge
work straight	=	work even
yarn over needle	=	yarn over
yarn round needle	=	yarn round
yarn forward	=	bring yarn to front of work
cast off	=	bind off
yarn back	=	bring yarn to back of work

Brightly coloured knitted socks make an ideal present for a country-lover.

KNITTING NEEDLE SIZES

Original UK	000	00	0	1	2	3	4	5
Metric (mm)	9	$8^1/_2$	8	$7^1/_2$	7	$6^1/_2$	6	$5^1/_2$
USA		15	13	–	11	10	10	9	8

Original UK	6	7	8	9	10	11	12	13	14
Metric (mm)	5	$4^1/_2$	4	$3^3/_4$	$3^1/_4$	3	$2^3/_4$	$2^1/_4$	2
					$3^1/_2$		$2^1/_2$		
USA		7	6	5	4	3	2	1	0	00

KNITTED WELLIE SOCKS ***

The perfect Christmas stocking, these wellie socks will be a blessing well into January snowstorms as well. Socks to wear inside wellington boots should be thick, firm, and roomy so that you can wear your usual socks underneath. These socks are worked in a wool-mix, aran-weight yarn, but the same pattern in double knitting yarn would given an everyday sock with the leg and foot length worked as required. Four-ply knitted to the same pattern would give socks for a child. To obtain a firm finish, these socks were knitted on needles two sizes smaller than those usually used for this yarn. You would also need to change needles for double knitting and four-ply yarn.

▲

Wendy Family Choice Aran × 50 gm balls:
red 5 balls; white 1 ball.
Set of four double-ended $3^3/4$ mm (No. 9)
knitting needles.

▼

Tension: 22 sts and 30 rows = 10 cm (4 in) in stocking stitch on $3^3/4$ mm (No. 9) needles.

Using red, cast on 66 sts, i.e., 22 sts on each of 3 needles, and work 3.5 cm ($1^1/2$ in) of K1, P1 rib, ending with a complete round.
Change to white and P1 round.
Work 2.5 cm (1 in) of K1, P 1 rib in white.
Change to red and P 1 round.
Cont in red, work a further 3.5 cm ($1^1/2$ in) in K1, P1 rib.

Change to stocking stitch (every round K) and work to a total length of 45 cm (18 in), ending with a complete round.

Shape heel:
K first 16 sts of round, then slip the last 17 sts of the round onto the other end of the same needle. Divide the remaining 33 sts onto 2 needles and leave for instep.
Return to the 33 heel sts and work:
Row 1 – slip 1, P to last st, K1.
Row 2 – slip 1, (K1, P1) to last 2, K2.
Repeat rows 1 and 2 eleven more times, then row 1 once more (25 pattern rows in all).
Row 26 – slip 1, K21, K2 tog, K1, turn.
Row 27 – slip 1, P12, P2 tog, P1, turn.
Row 28 – slip 1, K13, K2 tog, K1, turn.
Row 29 – slip 1, P14, P2 tog, P1, turn.
Row 30 – slip 1, K15, K2 tog, K1, turn.
Row 31 – slip 1, P16, P2 tog, P1, turn.
Cont in this way, working 1 more on each row before working 2 tog, until 23 of heel sts remain.
Next row – slip 1, K20, K2 tog, turn.
Next row – slip 1, P19, P2 tog, turn.
Next row – slip 1, K to end (21 sts).

Work foot:
Slip the 33 instep sts onto 1 needle. With right side facing and working onto needle containing the 21 sts, pick up and knit 15 sts evenly down the side of the heel; using a second needle, knit across the 33 instep sts; then, using a third needle, pick up and knit 15 stitches up the other side of the heel; knit 10 sts from the first needle (84 sts).
Re-arrange stitches:
Slip last 2 sts from instep needle onto end of third needle, and first 2 sts from instep needle onto end of first needle (28 sts on first needle, 29 sts on second, instep, needle, 27 sts on third needle).
Next round – first needle: K to last 3, K2 tog, K1; second needle: K to end; third needle: K1, K2 tog tbl, K to end.
Next round – K.
Repeat these 2 rounds until 58 sts remain.
Cont on these sts until work measures 23 cm (9 in) from back of heel, ending with a complete round.

Shape toe:
Next round – first needle: K to last 3, K2 tog, K1; second needle: K1, K2 tog tbl, K to last 3 sts, K2 tog, K1; third needle: K1, K2 tog tbl, K to end.
Next round – K.
Repeat these 2 rounds until 18 sts remain in total.
Next round – (K2 tog) to end (9 sts).
Break off yarn, thread through these 9 sts, draw up tightly and fasten off firmly.

Finished size: To fit most adults. Foot length 28 cm (11 in); foot circumference 25 cm (10 in); length from top of heel including the rib 45 cm (18 in)

An infallible way of keeping away witches is to hang a witch ball in a window by the door. A witch ball is a large silver blown-glass ball, like an oversize Christmas tree bauble. Witches perceive it as being full of a tangle of invisible threads which they must unravel before they can come into the house. As they are never able to achieve this before dawn, when they have to return home, they never have time to enter.

KNITTED TARTAN RUG *

This rug is knitted in a fashionable tartan, and is easy to make. The coloured stripes are knitted in and the checked stripe is embroidered.

▲

Wendy Ascot Chunky × 50 gm balls: plum 13 balls; black 2 balls; white 1 ball; yellow 1 ball.

▼

Tension: 14 sts and 20 rows = 10 cm (4 in) in stocking stitch on 6½ mm (No. 3) needles.

Note: Because the row is so long, use a circular needle from side to side, not in the round.

Make one main piece:
Using plum, cast on 158 sts and work stripe pattern in st st:
16 rows plum.
2 rows black.
1 row yellow.
5 rows plum.
1 row white.
8 rows plum.
1 row black.

Repeat these 34 rows 5 times, then work 16 further plum rows (186 rows in all).

Cast off loosely in plum.

To complete the rug, press according to ball-band instructions. Embroider vertical lines in any chosen random sequence to complete the tartan pattern, taking care to keep the work flat. The embroidered lines will be strong and clear if they are worked in a loose chain stitch, but if finer lines are preferred stem or backstitch could be used. Tartan stripes are arranged in a random group which is then repeated, with stripes of different colours close together, often followed by widely spaced stripes. Arrange the stripes in any way, using the photograph of the rug as a guide, or copy the sequence of the knitted stripes, echoing it in the other direction with embroidery. In order to keep the lines of embroidery straight, simply follow vertical lines of knitted stitches.

Fringe all four edges, matching the fringe pieces to the edge colour. To make the fringe, cut pieces of yarn about 13 cm (5 in) long. Take each piece of yarn and double it. Put a large crochet hook through the edge of the knitting, from wrong side to right side, about 5 mm (¼ in) in from the edge, hook the middle of the doubled piece of yarn and pull it through the knitting to form a loop on the wrong side. Thread the ends of the yarn through this loop and tighten to hold the yarn in place. Match the colour of the knots to the colour of the edge all round. When the fringe is complete trim any long ends to make it of an even length.

Finished size: approximately 113 × 93 cm (45 × 37 in)

ARAN CUSHIONS

Aran patterns are fascinating to knit and produce a thick texture which keeps out the Yorkshire Dale winds. However, they can be used for other items, such as cushion squares. Aran stitches make a comfortable, practical cushion and give you a chance to experiment with some of these beautiful textures, which you can then try on sweaters in the future. Size and tension are the same for all the cushions, and they are all finished and made up in the same way (see page 123). If you have difficulty following an Aran pattern, or in remembering your place it is a good idea to use a stick-on-peel-off memo note to mark the pattern row on the page and move it down as you work. Some people find that the short cable needles easily fall out of the stitches. To avoid this it is possible to buy cranked cable needles (from good craft or wool shops, which have a bend in the middle to keep the needle in place.

Another common problem with Aran knitting is that you lose your place in a row of stitches, especially if you are called away from the work in the middle of a row. Aran patterns are almost always written so that a number of stitches is worked in a certain way, and repeated to the end of the row. If small, separate slip knots are made in a contrastng coloured yarn, they can be placed on the needle between each pattern repeat and slipped from one needle to another as each row is worked, so that they continue to divide each group of stitches.

Aran knitting is much easier than it looks. If you feel daunted by the pattern, follow the instructions very carefully, and the pattern will work out. **Tension:** 19 sts and 24 rows = 10 cm (4 in) in stocking stitch on 4½ mm (No. 7) needles. Chosen needles are used throughout.

SPECIAL ABBREVIATIONS

C2B or C2F = cross 2 front or cross 2 back, i.e., knit into back (or front) of 2nd st on needle, then knit first st, slipping both off the needle at the same time.

C4B or C4F = cable 4 back or cable 4 front, i.e., slip next 2 sts onto a cable needle and hold at back (or front) of work, knit next 2 sts from lefthand needle, then knit sts from cable needle.

C6B or C6F = cable 6 back or cable 6 front, i.e., slip next 3 sts onto cable needle and hold at back (or front) of work, knit next 3 sts from left hand needle, then knit sts from cable needle.

T2F = twist 2 front, i.e., slip next st onto cable needle and hold at front of work, purl next st from lefthand needle, then knit st from cable needle.

T2B = twist 2 back, i.e., slip next st onto cable needle and hold at back of work, knit next st from lefthand needle, then purl st from cable needle.

T3F = twist 3 front, i.e., slip next 2 sts onto cable needle and hold at front of work, purl next st from lefthand needle, then knits sts from cable needle.

T3B = twist 3 back, i.e., slip next st onto cable needle and hold at back of work, knit next 2 sts from lefthand needle, then purl st from cable needle.

T5F = twist 5 front, i.e., slip next 3 sts onto cable needle and hold at front of work, purl next 2 sts from lefthand needle, then knit sts from cable needle.

T5B = twist 5 back, i.e., slip next 2 sts onto cable needle and hold at back of work, knit next 3 sts from lefthand needle, then purl sts from cable needle.

MB = make bobble, i.e., knit into front, back and front of next st, turn and K3, turn and P3, turn and K3, turn and sl 1, K2 tog, psso (bobble completed).

Finished size: approximately 40 cm (16 in) square.

Cushion covers provide a good introduction to using traditional Aran patterns. Displayed together, the variety of patterns and textures that are characteristic of this style of knitting are illustrated to full effect. From left to right: Bobble tree, Celtic plait, Lattice diamond and Medallion moss cable.

CREAM BOBBLE TREE CUSHION * * *
▲

Wendy Family Choice Aran in cream:
5 × 50 gm balls.
Cushion pad approximately 40 cm (16 in)
square, or equivalent filling.
▼

Make two pieces.
Cast on 90 sts.
The Bobble Tree pattern is worked over 18 sts and repeated 5 times across the work:
1st row (wrong side) – K8, P2, K8.
2nd row – P7, C2B, C2F, P7.
3rd row – K6, T2F, P2, T2B, K6.
4th row – P5, T2B, C2B, C2F, T2F, P5.
5th row – K4, T2F, K1, P4, K1, T2B, K4.
6th row – P3, T2B, P1, T2B, K2, T2F, P1, T2F, P3.
7th row – K3, P1, K2, P1, K1, P2, K1, P1, K2, P1, K3.
8th row – P3, MB, P1, T2B, P1, K2, P1, T2F, P1, MB, P3.
9th row – K5, P1, K2, P2, K2, P1, K5.
10th row – P5, MB, P2, K2, P2, MB, P5.
Repeat these 10 rows until work is square, ending with a 1st, 3rd, 5th or 7th row.
Cast off.

CREAM CELTIC PLAIT CUSHION * * *
▲

Wendy Family Choice Aran in cream: 7 ×
50 gm balls.
Cushion pad see above
▼

Make two pieces.
Cast on 83 sts.
Work increasing (wrong side) row – P1, (inc 1 st in next st in P) to end. (165 sts)
The pattern is worked over 10 sts, repeated 16 times (plus 5 to complete the pattern):
1st foundation row (right side) – K3, * P4, K6, rep from * to last 2 sts, P2.
2nd foundation row – K2, * P6, K4, rep from * to last 3 sts, P3.
1st row – K3, * P4, C6F, rep from * to last 2 sts, P2.

2nd row – K2, * P6, K4, rep from * to last 3 sts, P3.
3rd row – * T5F, T5B, rep from * to last 5 sts, T5F.
4th row – P3, * K4, P6, rep from * to last 2 sts, K2.
5th row – P2, * C6B, P4, rep from * to last 3 sts, K3.
6th row – As 4th row.
7th row – * T5B, T5F, rep from * to last 5 sts, T5B.
8th row – As 2nd row.
Repeat the last 8 rows until work is square, ending with a wrong side row.
Next row – K1, (K2 tog) to end (83 sts).
Cast off.

BROWN LATTICE DIAMOND CUSHION * *
▲

Wendy Family Choice Aran in brown: 4 ×
50 gm balls.
Cushion pad see above
▼

Make two pieces.
Cast on 84 sts.
The pattern is worked over 12 sts and repeated 7 times across the work:
1st row (right side) – K2, P8, * C4F, P8, repeat from * to last 2 sts, K2.
2nd row – P2, K8, * P4, K8, repeat from * to last 2 sts, P2.
3rd row – K2, P8, * K4, P8, repeat from * to last 2 sts, K2.
4th row – As 2nd row.
5th row – As 1st row.
6th row – As 2nd row.
7th row – * T3F, P6, T3B, repeat from * to end.
8th row – K1, P2, K6, P2, * K2, P2, K6, P2, repeat from * to last st, K1.
9th row – P1, T3F, P4, T3B, * P2, T3F, P4, T3B, repeat from * to last st, P1.
10th row – K2, P2, * K4, P2, rep from * to last 2 sts, K2.
11th row – P2, T3F, P2, T3B, * P4, T3F, P2, T3B, repeat from * to last 2 sts, P2.

12th row – K3, P2, K2, P2, * K6, P2, K2, P2, repeat from * to last 3 sts, K3.

13th row – P3, T3F, T3B, * P6, T3F, T3B, repeat from * to last 3 sts, P3.

14th row – K4, P4, * K8, P4, repeat from * to last 4 sts, K4.

15th row – P4, C4F, * P8, C4F, repeat from * to last 4 sts, P4.

16th row – As 14th row.

17th row – P4, K4, * P8, K4, rep from * to last 4 sts, P4.

18th row – As 14th row.

19th row – As 15th row.

20th row – As 14th row.

21st row – P3, T3B, T3F, * P6, T3B, T3F, repeat from * to last 3 sts, P3.

22nd row – As 12th row.

23rd row – P2, T3B, P2, T3F, * P4, T3B, P2, T3F, repeat from * to last 2 sts, P2.

24th row – As 10th row.

25th row – P1, T3B, P4, T3F, * P2, T3B, P4, T3F, repeat from * to last st, P1.

26th row – As 8th row.

27th row – * T3B, P6, T3F, repeat from * to end.

28th row – As 2nd row.

Repeat these 28 rows until work is square, ending with a 6th or 20th row. Cast off.

BEIGE MEDALLION MOSS CABLE CUSHION * * *

▲

Wendy Family Choice Aran in beige:
6 × 50 gm balls.
Cushion pad see opposite

▼

Make two pieces.

Cast on 114 sts.

The pattern is worked over 19 sts and repeated 6 times across the work:

1st row (right side) – P3, K4, (P1, K1) 3 times, K3, P3.

2nd row – K3, P3, (K1, P1) 4 times, P2, K3.

Repeat the last 2 rows once more.

5th row – P3, C6F, K1, C6B, P3.

6th row – K3, P13, K3.

7th row – P3, K13, P3.

Repeat the last 2 rows twice more.

12th row – K3, P13, K3.

13th row – P3, C6B, K1, C6F, P3.

14th row – As 2nd row.

15th row – As 1st row.

16th row – as 2nd row.

Repeat these 16 rows until work is square.
Cast off.

To make up a cushion

Do not press the knitted cover pieces. Turn one piece through 90° so that the cast on and cast off edges of one piece will be attached to the side edges of the other piece. Seam the pieces together, inserting the cushion pad or filling before closing up the fourth side, which can be closed with a zip fastener.

Aran cushion details: top row, left, Bobble tree, right, Celtic plait; bottom row, left Lattice diamond, right, Medallion moss cable.

HEIRLOOM BEDSPREADS * * *

Knitted bedspreads are very warm and attractive, helping to keep out a cold Yorkshire night and looking especially good in a farmhouse bedroom. This pattern is easy and quick to make, and the pieces can be worked separately and assembled at the end. If you have always wanted to make an heirloom cotton cover, this is the one to try, because its simplicity makes it less daunting. You can omit the yellow stripes if you wish, or adapt them to a design of your own.

▲

Wendy Miami Cotton mix double knitting: cream 22 × 50 gm balls; yellow 11 × 50 gm balls.

▼

Tension: 22 sts and 31 rows = 10 cm (4 in) in stocking stitch on 4 mm (No. 8) needles.
Chosen needles are used throughout.

SPECIAL ABBREVIATIONS

C = cream, Y = yellow.
MBY = Make bobble in yellow: using yellow, K4 times into the next st, * turn and K4 *, repeat from * to * 4 more times, turn and K4 tog.
SM = slip marker from lefthand needle to righthand needle.

Make 48 squares.
To make one square:
Before beginning, make a small slip knot of contrast yarn to act as a marker.

Using C, cast on 2 sts.
Row 1 – K twice into the first st, place the marker slip knot onto the righthand needle, K rem st. (3 sts).
Row 2 – K twice into the first st, SM, K rem 2. (4 sts).
Row 3 – K1, K twice into next st, SM, MBY in next st, K rem st. (5 sts).
Row 4 – K1, K twice into next st, SM, K rem 3. (6 sts).
Cont this way, working twice into the st before the marker on every row, and so inc by 1 st on every row throughout rem of work. At the same

time work in the following colour sequence.
Cont in C until sides of work measure 9 cm (3½ in), ending with a wrong-side row (the bobble made earlier is on the right side).
Change to Y and cont until sides of work measure 11 cm (4½ in) ending with a wrong-side row.
Change to C and cont until sides of work measure 18 cm (7 in) ending with a wrong-side row.
Change to Y and work rem of work in stripes of 6 rows Y, 6 rows C until sides of work measure approximately 25 cm (10 in) ending with a 6 row stripe of C. Cast off loosely.
Note: Using this first square as a master, ensure that all subsequent squares are exactly the same.

Do not press the squares. To assemble the bedspread, join the squares into 12 groups of 4 squares with the bobbles in the centre and matching the stripes. Join these 12 larger squares into 3 rows of 4 to give a final rectangle measuring approximately 150 × 200 cm (60 × 80 in).

EDGING

Using cream, and with 4 mm (No. 8) needles, cast on 11 sts.
Row 1 – K4, yrn, P2 tog, K4, yon, inc 1 st in last st.
Row 2 – K3, yrn, P2 tog, K4, yrn, P2 tog, K1, sl 1.
Row 3 – K4, yrn, P2 tog, K1, P2 tog, yon, K4.
Row 4 – K5, yrn, P2 tog, K2, yrn, P2 tog, K1, sl 1.
Row 5 – K4, yrn, P2 tog, K2, yrn, P2 tog, K3.
Row 6 – Cast off 3 (1 st on needle after cast off) yf, K5, yrn, P2 tog, K1, sl 1.
Repeat these 6 rows to form the pattern.
Cont in pattern until edging fits all around edge of bedspread. Leave stitches on a holder.

Sew the edging evenly around the whole edge, attaching it by its straight edge and gathering it around the bedspread corners so that the finished work will lie flat. Adjust the final length of the edging before casting it off and joining the ends together.

Finished size: Each square will have 25 cm (10 in) sides. A total of 48 squares will give a bedspread of approximately 200 × 150 cm (80 × 60 in) (excluding edging), suitable for a single bed.

A hand-knitted heirloom bedspread will be treasured and handed down from one generation to the next.

Index

ACKNOWLEDGEMENTS

The author and publishers would like to thank
the following suppliers who kindly supported this
book with materials.

·

For the knitting yarns:
Carter & Parker (Wendy Wools) Ltd,
Guiseley,
West Yorkshire LS20 9PD.
To find your nearest supplier, contact Customer
Services (0943) 872264.

·

For the flower presses:
Reeves Dryad presses are obtainable from all
good craft shops. In the event of any difficulty,
please write for a list of stockists to:
Reeves Dryad,
PO Box 38,
Leicester LE1 9BU.

·

For the tapestry and embroidery threads:
Coats Leisure Crafts Group Ltd,
39 Durham Street,
Glasgow G41 1BS.

Thanks are also due to the following, for their
help with the photography.

·

Food stylist
Jackie Hine

·

Locations
Trudie Proctor,
The Old Rectory,
Bletchingley,
Surrey.

·

Fresh flower basket (p. 25)
The Lemon Tree,
117 St John's Hill
London SW1

·

Stencilled items (pp. 46–7)
Elizabeth Keevill